名家笔下的迪奥
WRITINGS ON DIOR

3

克里斯汀·迪奥，梦之设计师
CHRISTIAN DIOR DESIGNER OF DREAMS

上海人民美術出版社 Artron Books 雅昌艺术图书

图书在版编目（C I P）数据

克里斯汀 · 迪奥，梦之设计师 / 巴黎克里斯汀 · 迪奥公司编 ; mot.tiff inside 工作室译 . -- 上海 : 上海人民美术出版社 , 2020.6

ISBN 978-7-5586-1388-3

Ⅰ . ①克… Ⅱ . ①巴… ② m… Ⅲ . ①服装设计－作品集－法国－现代 Ⅳ . ① TS941.28

中国版本图书馆 CIP 数据核字 (2020) 第 107856 号

克里斯汀 · 迪奥，梦之设计师

编　　者：巴黎克里斯汀 · 迪奥公司
翻　　译：mot.tiff inside 工作室
监　　制：包晨晖
责任编辑：郑舒佳
书籍装帧：CASSI EDITION
出版发行：上海人民美術出版社
社　　址：上海市长乐路 672 弄 33 号
印　　刷：雅昌文化（集团）有限公司
开　　本：787×1092 1/16
印　　张：21
版　　次：2020 年 6 月第 1 版
印　　次：2020 年 6 月第 1 次印刷
书　　号：ISBN 978-7-5586-1388-3
定　　价：347.00 元（全三册）

目录

导言：克里斯汀·迪奥的“全新梦风貌”

撰稿：奥丽悦·库仑

本次展览“克里斯汀·迪奥，梦之设计师”为迪奥时装屋描绘出了发展的轨迹，品牌从 1946 年开业直到今天始终是世界领先的时装屋之一。展览通过迪奥典藏馆中的时装、文件和图像收藏以及来自世界各地的博物馆、美术馆和私人收藏家的珍品，栩栩如生地展示了时装屋丰富有趣的历史，介绍了品牌创始人克里斯汀·迪奥的生平和创作，也纪念了其后的创意总监们。展览的举办以及精选评论文章的刊发，就是对迪奥时装屋七十余年来高级订制时装创作精髓的最好展示。

克里斯汀·迪奥时装屋的故事可以追溯到巴黎蒙田大道 30 号，这幢风格优雅的典型巴黎城市建筑至今仍是迪奥品牌的大本营。克里斯汀·迪奥在自传中提到，他年轻时想当建筑师，所以对这栋豪宅富有个性的建筑十分着迷，对其“规整紧凑的外观以及简洁优雅、毫无浮夸的气质”[1]赞赏有加。当纺织业大亨马塞尔·布萨克向迪奥先生提议资助其成立同名高级时装屋时，迪奥先生便租下了他喜爱的建筑，陶醉在翻新工程之中。他的密友、室内设计师维克多·格朗皮埃尔操刀装饰，与迪奥先生携手努力，拿出了“全是白色和珠灰，看上去很巴黎”[2]的主导色调。1947 年 2 月 12 日，时尚媒体编辑、买手和潜在客户蜂拥而至，在此亲眼目睹了时尚史上的这一关键时刻，克里斯汀·迪奥首秀，也就是后来被称为“新风貌”系列的发布。

菊丝婷·皮卡迪在关于迪奥的浪漫晚礼服（29 页）评论文章中指出，“新风貌”带来的影响之所以如此之强，是因为它所处的年代，那是 1939 年至 1945 年第二次世界大战结束之后，战争给人民带来了无比的艰辛和生活质量的压缩。战争初期，时装行业是法国最重要的产业之一，约有 30 万人从事高级时装制作和相关工作。德军占领时期，许多高级时装屋歇业，高级订制时装行业工会会长吕西安·勒龙经过努力，挫败了德国当局将时装工业转移到柏林的计划。战争结束后，时装贸易大量减少，也失去了海外顾客，尤其是美国这个最重要的市场。克里斯汀・迪奥于1946年底创立高级时装屋，时尚界都注视着巴黎，期待它重获霸主地位，再次成为时尚圈的中心。

在这样的大环境中，迪奥首秀开始前，媒体自然充满了期待和评论。其时，迪奥先生已经在时尚界小有名气，他曾在勒龙公司担任设计师，也曾与设计师皮埃尔・巴曼合作。1946年夏成立自己品牌的消息传出，业界的兴奋与日俱增。《时尚芭莎》总编辑卡梅尔・斯诺是迪奥创作的忠实拥趸，她与《Vogue》杂志美国版的主编贝蒂娜・巴拉德都是设计师的挚友，很早就开始在美国媒体和买手圈内推广迪奥品牌的时装。

首秀当天，争先恐后涌入蒙田大道30号的人群表现出极大的热情。发布秀开辟了先河，秀场里弥漫着品牌第一款香氛Miss Dior迪奥小姐香水的芬芳，卡特琳・泽彤（13页）深入探讨了这种走秀方式，模特信心满满，大步穿过秀场，全场观众无不惊叹。根据《时尚芭莎》的编辑安妮・斯科特・詹姆斯的回忆，模特"裙摆飞扬，从观众面前快步走过，对着空气微笑，摆出若无其事的表情。随着大裙摆旋转，百褶飘荡，塔夫绸沙沙作响，刺绣闪闪发光，各种色彩融为一体"。[3]迪奥设计的这些奢华服饰和"Corolle（花冠）"系列和"En 8"系列不负众望，真正捕捉到了所有人的兴奋点。"Corolle（花冠）"系列用抹胸打造上下颠倒的花朵形状设计，突出窄而溜的肩和细腰，臀部由喇叭裙紧紧包裹，下摆散开，远远超出当时流行的长度。这些礼服裙有些是使用了13米以上的布料制成，对仍生活在挥之不去的残酷战争阴影中的不少观众来说，这种肆意的奢华令他们精神为之一振。"En 8"系列又有不同之处，紧身裙贴在腿部，腰部也很贴身，构成数字"8"的轮廓。这个系列令时尚界雀跃，卡梅尔・斯诺将其命名为"新风貌"。

杰瑞・斯塔福德详细描述（17页）许多观众是如何为迪奥精美的色彩而感动，英国时尚记者维妮芙蕾德・刘易斯这样写道，迪奥先生拥有"超凡的用色天赋"，[4]许多杂志也详细介绍了他独特而有个性的色彩组合。《Elle》杂志写道，他使用蓝灰色、烟灰色、深灰色，就像天空在地面积水里的倒影，而米灰色则与杜勒花园铺着砂石的步道相呼应。[5]然而，"新风貌"的象征却是一件黑白相间的日装。它富有设计感的线条和黑白的配色非常适合在当时很少采用彩色印刷的报章上大批量复制。这套名为"Bar（迪奥套装）"的套装，由当时领导迪奥套装工坊的年轻裁缝皮尔・卡丹制作，灰白色的桑蚕丝绸外套裹住溜肩，腰部和臀部紧包，裙摆内精心加入衬垫使其散开。半身裙很长，由打细褶的黑色羊毛制成，配上一顶黑色大檐帽。《时装官方公报》对业界春季系列的报道就刊登有套装的插图，《Vogue》英国版还发布了迪奥先生的插画师密友克里斯汀・贝拉尔绘制的素描，称其为"具有茶杯曲线的紧身罗缎上衣"。[6]美国《时尚芭莎》则刊载了迪奥先生另一位好友插画家勒内・格鲁瓦的素描图，并指出"克里斯汀・迪奥设计的巴黎风下午日装'Bar（迪奥套装）'把握了新时尚的关键轮廓：钟形，天然罗缎上衣和百褶黑色长羊毛裙突出夸张的臀部线条"。[7]《Vogue》美国版上刊登塞尔日・巴尔金拍摄的整幅照片上，模特身着"Bar（迪奥套装）"侧身站在蒙田大道30号的楼梯上。"Bar（迪奥套装）"被时尚界精心挑选出来，成为"新风貌"的代言。

1955年，时装屋成立八年后，克里斯汀・迪奥在巴黎索邦大学演讲，身着品牌设计的模特也集体亮相，大师在阵容中没有忘记"Bar（迪奥套装）"。模特蕾内・布勒东穿着它，由摄影师威利・梅沃德的镜头记录下来，原始设计中的黑帽由天然染料染色的草帽代替。这张照片记录的形象后来成为套装的官方形象。自从1957年创始人迪奥先生突然仙逝以来，领导时装屋的六位创意总监的每一位都坚持将"Bar（迪奥套装）"视作品牌的重要标志，并且不断对其加以重新演绎。洪晃（25页）对迪奥文化遗产传承进行了梳理，展示每位创意总监如何对经典重新演绎。

“新风貌”在T台上的巨大成功与新时代女性主义的关系是充满争议的。毫无疑问，迪奥先生爱女性，他与妹妹卡特琳娜的关系很好，时装屋中许多重要职位都由女性同事担任，他得到了众多女同事的爱戴和拥护。他的得力助手蕾梦德·泽纳克尔就是女性员工，他称她为“第二个我”，是“坚定而有能力掌管所有业务”的人。[8] 玛格丽特·卡雷是工坊的技术总监，也是领导核心业务的女员工。娜迪亚·阿尔贝蒂尼对此进行了深入探讨（9页），迪奥品牌的经典轮廓和剪裁通过时装的创造而获得生命。迪奥先生的缪斯女神米萨·布里卡尔给他的不少作品带来灵感，同时也是女帽设计总监。然而，迪奥设计的轮廓也遭到过质疑，腰部紧身设计和长裙使人想起了从前女性受到压迫的时代，因为现在的人已经习惯了战争年代带给女性的自由，战时女性常着裤装和短而实用的裙子以及制服款服装。《不列颠尼亚和夏娃》杂志评论了新时尚的拥趸和抗议者之间的这场争执，并指出，越是对此大加宣传，越“可能会让更长的裙子和紧腰身受到更大欢迎，因为女性对新时尚的意识觉醒了，之后就要把它穿在身上”。[9] 女性需要“新风貌”。克里斯汀·迪奥的记者朋友艾涅斯婷·卡特回顾自己在时装界的漫长职业生涯时指出，对于她来说，“没有什么能比得上最初的迪奥时装系列那么令人振奋，从来没有什么能像它们这样成为普遍而经久不衰的时尚。女性无论高矮胖瘦，年老年少，“新风貌”都适合她们。[10]

如今，迪奥的创意总监玛丽亚·嘉茜娅·蔻丽就把触及和吸引所有女性放在自己设计的核心地位。她是时装屋第一位女性掌门人，她展示了将女性主义和女性气质完美结合的可能性，出自她手的系列从众多女性主义作家、艺术家及社会活动家的理论著作中获取灵感。本次展览展示了她诠释的“Bar（迪奥套装）”新版本，配以羽毛般轻盈的透明薄纱裙和经典套装外套，配上印有作家奇马曼达·南戈齐·阿迪奇的“我们都应该是女性主义者”字样的T恤，蔻丽不仅为克里斯汀·迪奥的原始设计引入了轻盈和现代的元素，也不忘迪奥先生的初心，为新一代顾客打造全新的“新风貌”设计。

[1] 克里斯汀·迪奥：《迪奥自传》，维多利亚和阿尔伯特博物馆出版社，2012年：10页。
[2] 同上：21页。
[3] 安妮·斯科特·詹姆斯，《貂蝉满座》，纽约达顿出版社，1952年：120页。
[4] 维妮芙蕾德·刘易斯，《闲谈者和旁观者》，1947年3月26日号：382页。
[5] 《Elle》，1947年3月4日号：9页。
[6] 《Vogue》英国版，1947年4月号：50页。
[7] 《时尚芭莎》，1947年5月号：130页。
[8] 克里斯汀·迪奥：《迪奥自传》，安托尼亚·弗莱泽译，伦敦魏登菲尔德和尼科尔森出版社，1957年：12页。
[9] 维妮芙蕾德·刘易斯，《不列颠尼亚和夏娃》。
[10] 艾涅斯婷·卡特，《内行谈时尚》，伦敦迈克尔·约瑟夫出版社，1974年：76页。

迪奥工坊的技艺

撰稿：娜迪亚·阿尔贝蒂尼

20 世纪 80 年代以来，迪奥时装屋就建立了品牌专属的文化遗产档案库，对于历任艺术总监来说，堪称是取之不竭的灵感来源。他们都会定期查阅这些珍贵资料。可以说品牌的基因库都在其中，保存了以各种线条、面料和色彩打造的数百个轮廓。从历史性设计模板获得启发是难得的好机会，可以更好地了解时装屋的过往和高级订制时装的制作工艺。

迪奥先生为 1949 年春夏高级订制系列“Trompe l'œil（幻象感）”设计了一件晚礼服，称为“Miss Dior（迪奥小姐）”，是向 1947 年推出的品牌第一款香氛致敬。香水的主调是优雅的西普调和独特的绿色香调，而同名花艺装饰由怒放的花朵组成，由朱迪思·巴比耶花艺工坊打造。从塞莉娅·贝尔坦在《高级订制时装，这片未知的土地》[1] 中对花艺行业进行的调查中可知，这个花艺工坊是为数不多的直到 20 世纪 50 年代仍然在巴黎营业的同类型花店之一，花店成立于 1845 年，位于多努街 18 号，由女创始人命名。花店在 1916 年发行的《巴黎优雅人士》杂志中也榜上有名，主要制造各种用人造花制成的装饰品。1928 年由时尚大师保罗·布瓦莱编辑的《巴黎奢侈品名录》上有花店刊登的广告版画，介绍其业务范围包括刺绣、发艺、冠冕和胸针等方面的设计。工坊每年为时装设计师制作六千多种花艺作品。自本世纪初以来，工坊的创作手法一直未有改变，先用巨大的木框把面料撑开，等其干燥后，再折叠起来，然后使用巨型机械压力机切割，切花朵的模板都是传统的，形状多样，有单个的花瓣，也有整朵花、叶、枝等等，全面满足设计师的需求。

“Miss Dior（迪奥小姐）”这条礼服裙上，由工坊的巧手裁缝们精心刺绣的花样覆盖了整条裙子，完美地模仿了繁花似锦的景色。工作台上摆放数十种花瓣，大小、形状和颜色各不相同，当它们从 19 世纪传承下来的压花机中成型出来后，这些纤细的小块面料仿佛拥有了生命。巧手们以手工一个一个地处理花瓣，先把球形的金属模具用明火加热，利用其给花瓣定型，造出或大或小的球形花苞，然后将其放在事先打湿的柔软平面上，稍加压力，根据工具的大小和施加的压力不同，使每个花瓣弯曲或展开的程度有所不同。这类刺绣中就有被用在“Miss Dior（迪奥小姐）”上的某些花样，有花瓣边缘不规则的粉红康乃馨、娇媚纤巧的茉莉花和橙花等。雏菊、紫罗兰和迪奥先生的幸运花铃兰也在其中，为怒放丰美的花束画上圆满的句号。千姿百态的各种花朵在搭配中达到了完美的和谐。

1949 年，朱迪思・巴比耶花艺工坊努力再现着令迪奥先生迷恋的大自然。人工花卉这项传统手艺在巴黎的勒马里耶工坊继承下来并传承至今。工坊成立于 1880 年，是当今这种手艺的唯一守护者。70 年后，他们接到了与之前略有不同的使命。当然，他们的主要工作依然是再现大自然，再则是修复因自然因素而褪色的裙装。保存在档案库的时装都没有最初那么鲜艳了，有些颜色已被氧化。接到任务后，花艺师们都活跃起来，一拿到档案库借给工坊的礼服裙，就从各个角度和所有缝合处仔细观察，从而确定并记录人造花的品种、花蕊和籽的大小。他们先要打样、切割，然后定型并逐瓣固定。在迪奥公司刺绣总监利耶斯先生的敏锐双眼及精准见解指导下，巧手工匠们用茶作为染料擦拭绣花板，为了使呈现结果更佳，还需要轻轻压一压这些花费了无尽时光造出的花朵。

尽管高级订制时装是一门艺术，但它首先是众多男女工匠不断交流、倾情合作的故事。巴黎的诸多知名工匠亲手打造了克里斯汀・迪奥的梦想，可以举出名字的就有巴比耶、吉尼斯提、拉内尔、勒马里耶、勒萨日、梅特拉尔和维尔蒙等等。时装大师对手艺人怀有由衷的敬意，因为手艺丰富了他的创作。传统的专业技术和用料的珍贵，让礼服更显华丽。

1955 年 8 月 3 日迪奥先生在巴黎索邦大学大教室演讲，在台下 4000 余名观众面前，他表达了自己与花艺师、羽毛工艺师和刺绣师之间的紧密联系。

“不使用装饰品来进行点缀的想法已经过时了。种类繁多的装饰品层出不穷，纽扣、刺绣、缎带、花边等都能让我们清晰地表达要赋予礼服裙的想法和意愿。从前的人在胸甲上雕刻或者在大衣上绣一个图案，这个叫做纹章，但也是含有装饰效果的。现在这些纹章图案的象征价值或许随着时间流逝失去了，但它仍然是裙装的组成部分，并不是凭空添加上去的。而且，时尚中最重要的就是利用对比效果，因此采用与面料对比性强烈的装饰物，以其或多或少的持久性对裙装形成一个补充，就像看似脆弱的纽扣可以突出羊毛外套的厚重质感，以及花纹凸起的金银线刺绣或富有立体感的珍珠刺绣可以更好地衬托薄纱类面料的轻巧。”

绰号勒贝的刺绣工匠勒内・贝格的一件刺绣作品，为玛丽亚・嘉茜娅・蔻丽为品牌创作首个高级订制时装系列提供了灵感。当她参观格兰维尔迪奥博物馆时，看到了这件 1954 年的刺绣样品，是由勒内・贝格和妻子安德蕾领导的刺绣工坊制造的。勒贝堪称史上最伟大的刺绣师之一。60 多年来，他们夫妇俩领导下的工坊为众多最伟大的时装设计师打造了华美的装饰。克里斯汀・迪奥也是他们夫妇的密友，与他一样，他们俩对鲜花充满热爱，仅在 1947 年至 1957 年期间就接到过 250 份花朵订单。安德蕾对色彩的敏锐感觉以及勒内的丰富经验吸引迪奥先生不断与他们合作。两位手工艺人通过大师勾画的草图以及心目中“花样女子”的浪漫幻想，充分表达了构图的才华。而且，勒贝以其绣出的数百种花卉图案和充满 18 世纪审美风情的样板也给予迪奥先生许多灵感。

这个样板在 1988 年由勒贝的后人捐赠给博物馆，表现的是 20 世纪 50 年代典型的女性轮廓，上围丰满，腰部纤细，配以花冠状散开的裙摆，裙摆上饰有春天花草图案。玛丽亚・嘉茜娅・蔻丽马上被其吸引住了，并在 2017 年春夏系列中设计了类似的款式。勒贝刺绣风格的显著特征是将传统设计与对创新的不懈追求巧妙地融合在一起，从而能够将刺绣完美地融入轮廓。春季花草图案是由真丝线、棉线和多色酒椰纤维绣成的。出自他手的鸢尾、剑兰和雏菊都以十种不同的针脚绣出，丰富多样的工艺体现出独特的质感和立体感，由于他在针脚之中还藏着一些半圆的幻彩刺绣，整体上散发着淡雅的光芒，就像掩映在薄薄的晨曦之中。样板体现出巧夺天工的丰富技术手段，启发了萨弗拉内·科尔堂贝尔工坊制作“Essence d'Herbier（草木精华）”露肩连衣裙，现保存于迪奥典藏馆以及墨尔本维多利亚国家美术馆。玛丽亚·嘉茜娅·蔻丽对这种奢华工艺带有现代感的诠释与致意，在 2019 年 Miss Dior 迪奥小姐香水广告大片中得到淋漓精致的表现，娜塔丽·波特曼身着风格华丽的裙装，让各种美丽的野花灵动起来，极具现代感。

不仅是品牌的分包商创造了时装的魔力，其实，时装屋本身也是造梦工厂。裙装工坊和套装工坊的老资格员工弗洛伦斯·舍艾和娜黛热·格南向公众揭示了打样的奥秘。员工对迪奥先生遣词造句的尊重，诸如“Bar（迪奥套装）”“轻盈”这类充满内涵，对于品牌身份形成而至关重要的词，在这间高级订制时装的殿堂内永远如雷贯耳。

在迪奥工坊内，一百多名员工不停地忙碌着，他们身穿简单的白大褂，就像克里斯汀·迪奥在工作室里的装束。裙装工坊有 60 名员工，套装工坊有 40 名员工，包含了各个年龄段，经验最丰富的巧手工匠已经 58 岁了，最年轻的学徒才 17 岁。每个人都使用相同的工具，工具简单得令人惊叹，一只纯手工制作的皮革小袋挂在腰上或脖子上，都是员工亲手缝制的，里边放着缝衣针、别针、安全别针、剪刀、顶针和卷尺，全是基础而高效的工具。工坊主任弗洛伦斯·舍艾和娜黛热·格南把两个工坊内的一切都安排得井井有条，自 1982 年以来，她们就为迪奥工坊效力了，对每种面料、每个针脚、每种技艺都了如指掌。像在所有的手工艺工作室工作的工匠一样，她们利用自己的专业知识和整个团队的精湛手艺来实现每位艺术总监的想法。

这些年来，高级订制时装的技术原理其实一直都没有变过。无论是裁样锁边，还是包边，一切工序都是手工完成的，只有少数直缝可以用缝纫机代替。顾客首次试样并量好各种尺寸，就要为其打造一个木质假人模型。裁缝团队根据顾客身体对假人进行改造，增宽或收紧，直至调整到完美合体的尺寸，作为顾客订单的打样基础体型。在制作高级订制时装样衣时，每个阶段都需要极度细致和无限精准。样衣用白布制成，一旦检验通过，就将正式成为样衣。按照迪奥先生的初衷，“Miss Dior（迪奥小姐）”礼服裙是由钢托抹胸和宽大的长裙组成，分为上下两部分令穿着更舒适，行动起来更优雅，裙子由两种不同厚度的象牙色薄纱制成，极度轻盈透明的面料，可以方便地处理透明感，里边带有薄纱加粉米色欧根纱衬裙，赋予裙摆形状及优雅的风度。抹胸内衬三文鱼色塔夫绸，穿着舒适。尽管迪奥的所有礼服裙包含的缝纫技术极其复杂，但又都必须体现出轻盈柔媚的风姿，正如“Essence d'Herbier（草木精华）”露肩连衣裙所展示的那样。仅缝纫这两件礼服裙就需要 300 多个工时的辛劳，还没有算上刺绣这部分呢。达到完美的境界需要时间。要通过勤学苦练将眼手合二为一，发展出触摸的直觉，这就是高级订制时装永不外泄的秘密。

高级订制时装在秀场亮相，效果往往令人震惊，裙装仿佛来自天外，而它们其实是不同手艺工坊巧手裁缝凭借精湛技艺的集体创作。自 1947 年以来，迪奥时装屋与手工艺工匠之间就一直保持着坚不可摧的联系，而今日依然不断携手迎接新的挑战。代代传承的手艺将面料奇迹般地变成时装，为高级订制时装制造影响力至关重要。关于这种将人类才智上升到理论高度的特点，迪奥先生在回忆录中这样谈到：“高级订制时装是手工可以制造的最精巧的事物之一，人类双手具有不可替代的价值，因为它们赋予自己创造出来的事物的特性是机器生产无法带来的，那就是诗意和生命。”[2]

[1] 塞莉娅·贝尔坦：《高级订制时装，这片未知的土地》，巴黎阿谢特出版社，1956 年。

[2] 克里斯汀·迪奥：《时尚是如何制作的？》，载《费加罗报文学版》1957 年 6 月 8 日号。

克里斯汀·迪奥的芬芳花园

撰稿：卡特琳·泽彤

花园对于克里斯汀·迪奥来说是灵感源泉的不二之选，因为他从小就在格兰维尔的别墅受到熏陶。他对大自然的各种形状、色彩、气味充满迷恋，对植物学兴趣很高，特别喜欢与母亲一道在屋子周围的花园中照顾花花草草，可以待上好几个小时。“我从母亲那里继承了她对花卉的热爱，我最喜欢与植物和园丁在一起。”[1] 对花卉的感情也是连接他与家人的纽带，首先是与母亲，随后与妹妹卡特琳娜，他们都在卡里昂分享过种植的快乐。“那是我首次真正在乡村生活，而且我很快爱上了这种生活方式，在土地上缓慢而辛苦地劳作，四季更替，萌芽中蕴含的不可解的谜团，让我爱上了它。（略）当我发现自己真的骨子里有耕作的天赋时，我决定和妹妹一起耕种房子四周的那一小块土地。”[2] 他在这个地区生活的经历促使他于 1950 年在距此地不远处购入墨山城堡。乡村与土地紧密相连，随着季节更替而变化，而迪奥先生又热爱园艺，他在这里找到了宁静，在下一个系列设计出来之前他都会迫不及待地来此休憩充电。在这里他仿佛脱离了时间的控制，不再受到时尚的影响，远离城市的喧嚣，让他的思维更加活跃，灵感奔涌，从第一个系列创作之初就是如此。

迪奥先生与童年好友赛尔日·爱夫特尔 - 路易什于 1947 年时装屋创建伊始，就成立了克里斯汀·迪奥香水品牌，这位合作伙伴曾执掌科蒂香水化妆品公司，在业界享有盛誉。他们希望为迪奥创作一张香味名片，完美伴随“新风貌”系列首秀。“然后 Miss Dior 迪奥小姐香水就诞生了，仿佛萤火虫穿过普罗旺斯的夜空，绿茉莉与夜晚和土地的芬芳共同吟唱。”[3] 香水的命名是克里斯汀·迪奥和米萨·布里卡尔心有灵犀，一起想出来的。米萨·布里卡尔看见卡特琳娜·迪奥前来探班，于是呼唤迪奥先生：“快看，迪奥小姐来了！”他却回答道：“迪奥小姐！香水的名字找到了！”

Miss Dior 迪奥小姐香水应该“散发着爱的气息”，是香水史上最早的绿色西普调之一，出现时间比科蒂的西普香水和巴曼的绿风香水稍晚。它是由保罗·瓦谢调制的，香脂味和芳香味的前调是由玫瑰、栀子花、鸢尾、茉莉和水仙组成，中尾调中含有西普、龙涎和皮革味，香水浓度远高于平均水平，所以十分丰富，芬芳袭人，仿佛发布秀那天沙龙中的无数鲜花花束那么浓艳，也仿佛“Corolle（花冠）”系列礼服使用的数千米面料那么慷慨。

最初的香水瓶是由克里斯汀·迪奥亲手设计的，采用古希腊陶土双耳瓶的形状，这是古代用于保存易变质和珍贵食物的容器。这个形状也启发了 1947 年春夏系列“En 8”的轮廓。“礼服裙与香水瓶都带有同样的印记，并且双剑合璧。昂贵的小香水瓶纤纤细腰，礼服裙曲线分明。”[4]香水的包装采用千鸟格图案，克里斯汀·迪奥邀请插画师勒内·格鲁瓦绘制广告画。1952 年圣诞限量版瓶身是克里斯汀·迪奥爱犬鲍比的样子，由费尔南·凯利 - 克拉设计，小狗站在亭子形状的礼盒里特别可爱。

Diorissimo 香水由埃德蒙·鲁德尼兹卡于 1956 年调制，花香调的香水也成为品牌的象征，它仿佛一曲交响乐，借用了音乐的语言，获得最高等的赞誉，它的诞生是迪奥公司历史上的辉煌时刻。它的诞生源自克里斯汀·迪奥和埃德蒙·鲁德尼兹卡这两位创作大师的相遇和灵感的碰撞，堪称调香大师最成功的作品之一，通过极致的提炼终于打造出自己的风格和印记。

不仅是瓶中的香水，就连香水瓶都符合这两位天才人物的喜好，也配得上他们的精湛技艺。

这款香水是迪奥先生的最爱，因为它完美再现了他最喜欢的幸运花的香气。他上衣扣眼里永远插着一枝铃兰，也会将铃兰放在小首饰盒内藏在胸口的口袋里，铃兰的名字、形态、图案都被他用在珠宝、帽子、鞋履甚至礼服设计上，1954 年春夏就有整个系列叫做“铃兰”。

埃德蒙·鲁德尼兹卡也对铃兰有所痴迷，他在卡布里的花园中种满了铃兰，春天一来，就是芬芳满园，许多年来，在铃兰盛开的每一天，他都能感知它们香气的变化，体会不同的心情，并记录铃兰带给他的各种印象。

这种香气是带来幸运的芬芳，甜美、清新，象征春天的觉醒。它不是单一的香调，而是鲜花交融合奏，依兰与橙花翩翩起舞，茉莉和格拉斯玫瑰倾情对歌，唤醒无声的花朵，让它们高歌不止。

这支香水瓶也是古希腊陶土双耳瓶的形状，由克里斯汀·迪奥设计，由巴卡拉制造，瓶身上带有镀金铜塞，上面镶着一圈圈精美的金环，缠绕着一枝枝铃兰。这是克里斯汀·迪奥生前创作的最后一支香水。1957 年 10 月 29 日他下葬的时候，前来吊唁的时尚界人士在他的墓上盖满了铃兰。

1966 年，在埃德蒙·鲁德尼兹卡的手中，迪奥清新之水诞生了，“这款花香调的香水香气淡雅清新而持久，是青春力量的绝好象征”。[5]

它的简洁与清新的风格开辟了一类新型香水的调配方向，并且长盛不衰，各种清新之水应运而生，都属于古龙水，既适合男性也适合女性，带有柑橘香调。1953 年清新之水脱颖而出，在迪奥先生的心目中和皮肤上取代了薰衣草水的王者地位。这也是有史以来第一款配方中使用了人工合成茉莉花香味分子希蒂莺的香水，为花香调带来清新和力量。

1972 年，迪奥清新之水的淡雅清新的花香调也启发埃德蒙·鲁德尼兹卡调制出他最喜欢的一款香水，Diorella。这是他首次调配带有果香调的淡花香型香水，前调为西西里柠檬，中调呈现出金银花与桃子味的双重香气，后调为茉莉花，像水一样清澈透明。

由爱德华·弗莱西耶创造的迪奥奇葩淡香水为 20 世纪 80 年代带来了一场嗅觉震撼，花香调为主的香水以晚香玉为主调，也散发出玫瑰和康乃馨的香气，后调为带有李子味的果味香调和以八角茴香、肉桂为主的香料味，打造出甜美的花香调，蔚然成风，更成为经典。

让 - 路易·修萨克于 1991 年调配的迪奥沙丘淡香水给人带来的感觉，仿佛面对茫茫大海边一望无垠的沙滩，香辛料香调为以牡丹花香为主的水生调兼海洋调主调带来一丝温暖。

迪奥真我香水于 1999 年推出，是调香师卡利斯·贝克尔的杰作。其中的香味令人想起让·科克多对克里斯汀·迪奥的评论："他是我们这个时代的一位敏捷的天才，他神奇的名字是上帝与金子的结合。"[6] 这款香水的香调结合了花香与"满满一篮成熟的水果"香加上一丝浸泡在巴纽尔斯葡萄酒中的李子味。迪奥真我香水继承了 Miss Dior 迪奥小姐香水丰富肆意的热情，虽然不再与无穷无尽的面料搭配，但是它像一阵金色的倾盆大雨，以自己的方式为前十年时尚界和调香界的过度简约画上句号。

迪奥公司彼时已经将艺术总监的宝座让约翰·加里亚诺继承，才华横溢的加里亚诺不负时装屋的盛名，他设计的巴洛克风时装是时尚史的风向标，蕴含着成千上万种历史和各国民族精彩影响的戏剧化时装。

迪奥真我香水瓶设计采用克里斯汀·迪奥所钟爱的古希腊陶土双耳瓶形状，由埃尔维·范·德·斯特拉唐设计，并饰有约翰·加里亚诺创作初期使用过的非洲马赛部落项链。此后，弗朗索瓦·德马奇调配出迪奥真我纯香、迪奥真我倾世之金、迪奥真我心悦等各种广受欢迎的版本。

弗朗索瓦·德马奇从 2006 年春季开始担任迪奥公司的调香师，也是克里斯汀·迪奥品牌系列香水的创造者，系列中既包含身负悠久历史的数款香水，也有许多新作品，全部都采用格拉斯地区的最高质量花朵和各种高级原材料打造。他是公司丰富调香历史遗产的守护者，肩上担负着保持品牌身份连贯性的重任，同时也要为迪奥先生挚爱的花朵谱写全新的香气交响曲。富有内涵的蒙田大道灰香水更名为迪奥灰香水，却保留原有的香调。墨山城堡香水向迪奥先生心爱的妹妹卡特琳娜在卡里昂培育的洋蔷薇致敬。Lucky 香水是迪奥之韵淡香水开创的的幸运花香水系列，变身为迪奥先生最喜爱的花朵，"它是人们在每一个想祈求幸福的时刻都会用到的香水"。

迪奥香水从创始至今都由调香师根据格拉斯地区的香精植物创造，这些植物在这片沃土上被满怀关爱的双手种植下去，并在克里斯汀·迪奥、卡特琳娜·迪奥、弗朗索瓦·德马奇最爱的这片绝美风景中渐渐长大。如今香花园的种植者们成立了行业工会，努力保护珍贵的原材料及其精湛的传统工艺，联合国教科文组织也将其列为非物质文化遗产加以保护，使素馨茉莉、洋蔷薇、晚香玉、橙花、鸢尾、水仙和含羞草等迪奥先生珍爱的香气永不消逝。

迪奥品牌的调香历史在历代调香师手中传承不绝，由 Miss Dior 迪奥小姐和真我香水的花香调热烈奔放地赞誉，时而具体，时而抽象，但都是无比真诚而明朗。香水中蕴含了如此多承载着深爱的花朵，只要轻轻喷洒若干，就能马上置身于芬芳的花园中。

1 克里斯汀·迪奥：《迪奥自传》，维多利亚和阿尔伯特博物馆出版社，2012 年：168 页。
2 同上：186 页。
3 迪奥先生 1950 年左右访谈时说的话。
4 柯莱特关于时尚设计师品牌推出的香水，载《芬芳》，见《嗅觉歌剧》专辑一书前言，浪凡香水公司出版，1949 年。
5 埃德蒙·鲁德尼兹卡的话，载《克里斯汀·迪奥，另一个我》，1987 年。
6 文森·乐海：《克里斯汀·迪奥香水，香水 70 年》，载《克里斯汀·迪奥香氛之精神展览目录》，米兰希尔瓦纳出版社，2017 年 5 月：59 页。

斑斓世界

撰稿：杰瑞·斯塔福德

为什么时尚和女性不能是缤纷而动人的呢？——克里斯汀·迪奥
一抹色彩就能让您旧貌换新颜。——克里斯汀·迪奥

1947 年，美国抽象表现主义大师杰克逊·波洛克创作了富有开创性的第一幅“滴画”；金发碧眼的艳星玛丽莲·梦露首次亮相银幕；一个大陨石坠落在苏联腹地老爷岭，形成巨大的撞击坑。同年，克里斯汀·迪奥举办发布秀，以自己的剪裁和色彩发布充满革命性的时尚宣言，被称作“新风貌”，在世界上引起的震撼决不亚于前三者。首秀与时装系列为日后的设计定下基调，为时装重新定义，“服装是颂扬女性身体之美的瞬息建筑”。

所有艺术家，无论是设计师、画家还是作曲家，都深知色彩中孕育的力量、超能力以及对眼睛和灵魂不为人知的作用。他们明白，当我们看到颜色时，我们实际上感知的是一个相互作用的过程。色彩是承载着信息的语言，具有在社会生活中进行交流的能力，在充斥着各种色彩的自然世界中也是如此。

克里斯汀·迪奥创作生涯伊始就展示出丰富华美的各种色彩搭配，以及种种独特的色调运用，每一种色彩都与大师本人的经历息息相关，也对他本人及历任继任者意义非凡。

迪奥先生的首秀时装系列以及所有克里斯汀·迪奥设计的系列，都是对色彩的热情赞颂，大师让顾客们领略到一系列精心调配、高雅至极的色调，从娇嫩的白色、饱含大自然气息的粉色、怒放的黄色、泛青的釉绿色、神秘的蓝色、宁静的灰色，到优雅的黑色，还时不时用他最爱的“特拉法加”红来点缀，就像歌剧中悠扬的咏叹调一般，堪称画龙点睛之笔。

时装沙龙中，他最欣赏的“亲爱”的模特们在凝神摒息的观众眼前走秀，戴着手套的纤手会大大获得赞美、惊叹甚至震撼，如 1947-1948 年秋冬系列的“热烈的红”和“撒旦红”为“新

风貌”首秀注入活力，随着模特们在沙龙中的脚步，插画师勒内·格鲁瓦手中的羽毛笔笔走龙蛇，用鲜红墨水勾勒下她们的曼妙身姿，登上全世界的时尚杂志，成为经典。

设计师采用红色这种标志性的色彩，的确是出于热爱，他认为红色能够带来好运和魅力。似火红色在中国文化中也象征着吉祥和欢乐。

品牌现任设计师玛丽亚·嘉茜娅·蔻丽运用自己独有的色彩表达，续写着品牌的传统，像迪奥先生诠释的那样，采用“精准而含有寓意的色彩，对于今天仍然选择迪奥品牌设计的顾客而言，是可以传达重要讯息的”，她不忘初心，对完美红色的追求仍在继续。迪奥化妆品公司的创意总监彼得·菲利普斯认为，红色永远代表着“激情、爱、生命和迪奥”！

克里斯汀·迪奥不仅在时装设计上使用了丰富的色彩，而且还为经历过二战，对色彩极度渴求的顾客们创造出用色大胆极富开创性的化妆品系列。

1953 年 4 月，大师在故乡格兰维尔结交的发小赛尔日·爱夫特尔 - 路易什推出了“迪奥之红”口红，并大胆地推出从艳丽的鲜红色到纯净的橙色等八种色调的口红。

1968 年，迪奥先生请新上任的创意总监赛吉·卢坦斯为品牌创造了整个系列的彩妆，命名为“Explosion de Couleurs（迸发的色彩）”。全新系列包括多达 40 种不同色彩的口红。新系列的发布由知名摄影师吉·布尔丹拍摄广告大片，在片中可以看到卢坦斯如何在模特的脸上勾画出鲜艳生动的颜色，蓝色、绿色和黄色等原色，仿佛彩虹一般。此举标志着迪奥在化妆品与色彩领域掀起了另一场革命，正如彼得·菲利普斯强调，“迪奥就是色彩的艺术”！

克里斯汀·迪奥一直保持着对童年时代故乡海边小镇格兰维尔的回忆，他在父母的别墅 Les Rhumbs 的淡粉色石墙下成长，低垂的铅灰色天空、浪涛澎湃的大海、海岸线上露出的悬崖和岩石，都是他记忆中无法抹去的，激发他后来创造出淡雅怀旧的“grisaille（灰色调）”系列配色，大师很快就将其融入到品牌标志性的红色和诗意盎然的花卉图案中，灰色诚然是记忆中最珍贵的色彩，但有些单调了，需要补充。灰色是作为基调存在的，是让大师尽情添加亮色的画布。

尽管克里斯汀·迪奥声称自己的用色并没有受到特定的哲学思潮或美学理论的影响，但他对艺术史有丰富广泛的了解，并且对 19 世纪下半叶至 20 世纪初现代艺术运动对色彩运用方面产生的革命性变化深有体会，最令他印象深刻的是青骑士画派艺术家那种野兽派风格的作品，以及他心爱的印象派画家在诺曼底创作的众多室外写生画，画的正是他度过田园诗般的童年的家乡，梵高的后印象主义作品以及奥托·迪克斯和马克斯·贝克曼为代表的表现主义，以及布拉克、费尔南·莱热、毕加索等人的立体派抽象主义画作都是他的最爱。

迪奥先生的父母于 1905 年购买别墅并命名为“Les Rhumbs”。房子是由一位船主建造的，采取了当时流行的海滨浴场风格，墙壁“是温柔的粉色，带有灰色的砂石”。大师一生都对这两种颜色保持衷心的喜爱。

迪奥先生还特别欣赏 18 世纪洛可可风格画家尼古拉·兰克莱笔下的粉色色调，《跳舞的卡马尔歌小姐》（1730 年）和《林中舞会》（1722 年）等画作，完全唤起了花样女子的绽放之情，不仅在 18 世纪摄政时期的法国引起轰动，也让大师汲取灵感，从而将女性塑造成娇媚的花朵，打造“新风貌”的种种轮廓，“Corolle（花冠）”轮廓造型让裙裾像花瓣一样铺洒，为他创作的新系列带来耳目一新的感觉。当年时装秀的观众记忆中也充斥着有关“玫瑰花”“玫瑰红”“玫瑰粉”等一系列词汇。

对同色系内色彩纯度、明度等细微差别的精准描述都能说明时尚大师敏锐的色彩感觉和他对构图的深刻理解，这也是受到了诺曼底的童年经历的影响和启迪。迪奥先生利用色彩、印花和刺绣等种种手段，全方位开发了这片繁花遍地的风景的所有可能性。

粉色一直激发着迪奥公司的设计灵感，也被赋予了全新的意义，尤其是对于玛丽亚·嘉茜娅·蔻丽而言，她口中的色彩是“现代女性主义的新堡垒，歌颂并推行多样性的价值观，并在接受刻板印象的同时，发挥其优势，为自己的目标所用”。

反观迪奥先生的一生，也能够称为“玫瑰色的人生”，就像法国香颂女王伊迪丝·琵雅芙歌中所唱那样。伊迪丝·琵雅芙最爱的也是迪奥品牌的时装和口红。1955 年，她来到大师的沙龙时，玫瑰粉色已经成为品牌的标志性色彩，与“迪奥之红”那种性感大红一样具有辨识力，同样，大师在沙龙内部装饰也采用了标志性的色调，他称之为“我最爱的路易十六风格，特里亚农灰色和白色”。迪奥先生甫一使用这些颜色，就立即引起了顾客的兴趣和媒体的关注。世界上很少有时尚设计师能够既有效又持久地使用某些色彩，来让自己的作品有辨识度。在“新风貌”系列及其在奢侈品时尚领域掀起的前所未有的革命中，这些色彩是获得胜利的代名词。

迪奥先生不仅为品牌设计了所有标志性的色彩，他也在 1954 年撰写的《时尚小词典》中赞扬了灰色的优点，称其为“最方便实用且优雅的中性色”。他还觉得灰色是巴黎的颜色，他在时装秀上不断推出千变万化的各种灰色，配上模特的各种造型，在秀场笔记中被称作“城墙灰”“斑点蛾灰”“叹息灰”“珠光亮的天空色”“晨曦灰”和“冬季灰”等等。

白色与灰色搭配，的确是位于蒙田大道 30 号的迪奥公司总部主导的色彩，也是迪奥先生在巴黎远郊米利拉福雷的乡间度假胜地古德莱磨坊的“法国外省”风格的室内装修主色调。1954 年春夏系列是迪奥先生为铃兰花献上的礼赞，这种象征纯真无瑕的花经常与圣母玛利亚联系在一起，他以此为名创作了“Muguet（铃兰）”系列，表现自己的喜爱之情。勒内·格鲁瓦为迪奥推出的第一款香氛 Miss Dior 迪奥小姐香水创作插图，他用寥寥几笔勾画出的优雅天鹅也是白色的。在迪奥工坊中的扶手椅都饰以白色的靠垫，模特身披七彩布料轻盈地穿过沙龙，令好奇的眼睛无从窥探新设计的奥秘。如今，玛丽亚·嘉茜娅·蔻丽将色彩“与科学带来的乐观主义和对现实进行的实证主义方法研究联系起来”。

克里斯汀·迪奥平时在日常场合喜欢穿白色，黑色是他在公开场合正式露面时选择的颜色。黑色是风流倜傥的高雅色彩，也是大师最喜欢的 1920 年代德国表现主义电影的主导色调，比如罗伯特·威恩 1922 年执导的《卡里加里博士的小屋》的黑色调，以及路易丝·布鲁克斯在格奥尔格·威廉·帕布斯特 1929 年的大作《潘多拉魔盒》中角色头发的鸦翅黑。法国女歌手朱丽叶·格雷科被迪奥先生称作“穿着烟管裤和地洞老鼠那么黑的黑色毛衣的一代青年的完美代表”，而伊迪丝·琵雅芙也表现出了那种简练的高雅。黑色是让·科克多线条鲜明感性的新古典主义素描的色调，也是克里斯汀·迪奥在他的乡间别墅墨山城堡自己创作的那些小版画的色调，城堡四周环绕着柏树覆盖的层层山峦的黑色剪影，为他营造了安宁与庇护。

对于克里斯汀·迪奥来说，黑色不能脱离白色而存在，就像阴阳的辩证关系，黑与白的对比无处不在，在洋装和裙子的反差中，也在标志性设计千鸟格纹主题中，或织入经纬，或印在配饰和包装上。

克里斯汀·迪奥在《时尚小词典》中称黑色为“所有颜色中最受欢迎、最好搭配、最优雅的一种”。黑色是时尚大师永远不会枯竭的灵感源泉，与绘画大师马蒂斯一样，黑色的用途是展现“力量，像压舱物那样”，黑色最能够体现简单、率真且与众不同的轮廓，轮廓也能够与色调互相呼应，例如“Dame en Noir（黑衣贵妇）”（1947-1948 年秋冬系列）、“Panthère Noire（黑豹）”（1952-1953 年秋冬系列）以及“Collet Noir（黑衣领）”（1950-1951 年秋冬系列）等。黑色是 1955-1956 年秋冬高级订制“Soirée de Paris（巴黎之夜）”礼服轮廓的色调，乌黑头发的名模多维玛在 1955 年由摄影大师理查德·阿维顿拍下的照片中留下了不朽的影像，长着一对黑色眼睛，厚嘴唇的名模维克多娃·杜特罗像位黑色天使，她的身影在蒙田大道 30 号的灰白色调秀场中亦久久不能磨灭。

对于迪奥这位时尚大师而言，黑色的唯一强大竞争对手是海军蓝，因为这两种颜色都带有相类似的一丝神秘。

迪奥蓝是优雅神秘的蓝色，是 1949 年创作的“Junon（朱诺）”礼服裙的蓝色，超大裙摆的灰色薄纱长裙搭配大领口，上边点缀着许多深蓝色绣上去的亮片。蓝色也是“蓝色顾客”最爱的颜色，她是来自美国罗得岛州纽波特的一位菲尔士通夫人，只购买迪奥设计的皇室蓝颜色的礼服裙。蓝色也是凡尔赛宫和玛丽·安东奈特服饰上体现出来的精美蓝色，迪奥先生在 1955 年春夏高级订制系列中对此大加赞颂。蓝色也是对 18 世纪这个年代的怀念与理想化表现，对当时纸醉金迷的舞会、歌剧表演的怀念，因为法国大革命之前，所有的椅子和挂毯都是天堂般的蓝色。

克里斯汀·迪奥不仅可以像他的名作《时尚小词典》中所建议的那样，专门为黑色写上整整一本专著，还可以为所有的色彩洋洋洒洒地写上好几本书，因为他对这些主题实在太敏感了。他不仅在童年时代就对多彩的大自然产生了赞叹之情，而且他对同时代的许多前卫艺术家也有着深刻的认识与了解，从而获得了敏锐的色彩洞察力。

1927 年，虽然彼时的迪奥先生还是一个年轻小伙，但他已经是巴黎文艺界艺术家、作家和美学家们的圈内人，并已经遇见了与之称兄道弟的密友画家兼舞台设计师克里斯汀·贝拉尔、诗人兼电影制片人让·科克多、文艺界名人玛丽·洛尔·德·诺瓦耶伯爵夫人等，迪奥先生与朋友雅克·蓬让合伙在巴黎市中心博埃希街创办画廊，为 20 世纪上半叶许多最有影响力的著名艺术家做过展览，包括基里科、乌特里洛、达利、杜菲、玛丽·罗兰珊、莱热、让·里尔萨、毕加索，扎德金、布拉克、马约尔等人。迪奥与这些艺术家的友谊和了解对于他自己对色彩的探索以及他设计的作品都至关重要。

玛丽亚·嘉茜娅·蔻丽在今天的创作中继续弘扬着这种精神，2018 春夏成衣系列中，她从法国女艺术家妮基·桑法勒色彩鲜艳的作品出发，以白色为底板，创造出她自己形容为“带有条纹或波点等图案，是迪奥设计美学全新性格和口号”的作品。

在她设计的 2019 年早秋高级订制系列中，还使用了女画家索尼娅·德劳内喜欢用的那种水彩晕染色调，引起满座惊奇，用她的话说，这种色彩“对我和时装屋来说，都是超越舒适圈的尝试”。

各种色彩及其肉眼可见的渐变色系依然对艺术家和科学家的思想和眼睛产生着独特的吸引力，最早的这些艺术家和科学家是七万五千年前的穴居人，在岩洞的壁上用木炭和红色岩石等矿物作画笔涂抹，画出各种原始人的符号和象征图案；而当今的纳米技术专家、数码世界探险家和玛丽亚·嘉茜娅·蔻丽和彼得·菲利普斯等设计大师也继续感受到色彩的吸引力。

他们对色彩艺术及语言的创新，预示着迪奥品牌和迪奥“花样女子”新时代的到来，在不断发展变化之中永远蜕变，越来越美。

迪奥：将“新风貌”带给世界的环球旅行家

撰稿：杨忆非

迪奥先生的父母曾希望他成为一名外交官，但出于对艺术和时尚的无比热情，克里斯汀·迪奥最终成为史上最伟大的设计师之一。然而，他仍然周游世界，从欧洲到南美洲，用其惊人的创造力征服了世界。无疑，他已成为法兰西风格的首席大使。

旅行精神根植于迪奥先生的内心深处。英国是他年轻时访问的第一个国家。在20多岁的时候，他的旅途从雅典到列宁格勒和伊斯坦布尔，还在地中海巴利阿里群岛待了一年。自他推出自己的处女作以来，他就致力于从所观察到的文化中汲取灵感，并以那些地方的名字来命名他的系列，来向旅途中的建筑、景观和艺术致敬。

随着他声名鹊起，他的时装业务也扩展到全球领域，从美国到日本，从墨西哥到澳大利亚，成为了真正的时尚帝国。他在世界各地开设精品店，并授权许可他的设计，这带给他更多的风格启发，也有更多机会进行旅行。他深受世界各地顾客的喜爱，是最早认识到全球市场重要性的国际设计师之一。

迪奥先生不仅仅通过脚步，还有他的丰富想象力和再创造将他的创意和旅行联系起来，这些精神和手法后来也被他的继任者继承了下来。通过他们的精彩诠释，全球各地的多元文化踏入了国际时尚的殿堂。

远东一直是迪奥先生的创意乐土。他在诺曼底的格兰维尔长大，家里的装饰充满了日本风情。“喜多川歌磨和葛饰北斋所作的浮世绘作品装饰着我的西斯廷教堂，我常常驻足膜拜，一看就是几个小时……”他在回忆录中回忆道。那些精致的真丝刺绣给他留下了难以磨灭的回忆，此后他常常在自己的设计中进行回味。他总爱选择带有传统日本风格图案的丝绸进行创作。1953年，他打造出一款名为“Jardin Japonais（日本花园）”的套装，装点着细巧的樱花和花鸟图案。1954年，他为芭蕾舞演员玛格特·芳廷创作了一款穿着在芭蕾舞剧《日本亮相》的演出服，以和服为灵感启发。他还派时装屋的模特在东京展示其高级订制时装，还为日本美智子皇后的民事婚礼设计了三套礼服。

在他的自传中，迪奥还经常不厌其烦地描述亚洲工艺和设计的复杂性，以及他对印度和日本的面料和设计的喜爱，此后，他致力于这些东方的传统融入他的现代美学。他深深地被印度的装饰技术所吸引，并将传统刺绣加入到他的晚礼服中，这使他的礼服艳冠群芳，其本人也成为了有史以来最伟大的礼服设计师之一。模特安妮・冈宁和芭芭拉・马伦穿着他的设计漫步在印度迈索尔和新德里，被诺曼・帕金森拍摄下来，登上了英国版《Vogue》，展开了一幅融合东西方魅力的美丽图卷。

对迪奥先生来说，中国是另一处永恒的魅力的所在，尽管这位时装设计师一生中从未去过那里。在他的首个系列中，除了著名的“新风貌”，他还设计了一件上海风格启发的大衣，并配东京风格的帽饰。随后，他创作了“北京”“上海”“中国风情”“中国之夜”“香港”“中国蓝天”等款式，向这个古老而神秘的国度致敬。中国书法和传统旗袍也得到了他的青睐。1950 年，他将这些表意符号用于西式连衣裙上的印花，1955 年秋冬季，他则用鲜艳的粉红色重新诠释旗袍。他创作的中国式长袍被《Vogue》杂志点名称赞，杂志表示：“设计师选择了东方作为灵感主题，但实际上打造出了西方的当代的设计。”

对迪奥先生来说：在太平洋另一边的美洲，尤其是中美洲和南美洲，是另一个充满激情的征途。1948 年，他前往美国，同年，内曼・马库斯百货在达拉斯为他颁发了时尚大奖。这位设计师随后开始了美洲大陆的胜利之旅。1953 年，他在委内瑞拉的加拉加斯受到了热烈的欢迎。第二年，他在波哥大、墨西哥城、巴拿马城、波多黎各、金斯敦、库拉岛和哈瓦那举行了一场巡回时装秀。那时候，他经常光顾位于墨西哥城中心地带的哈凡尼豪华的精品店魅力公司和百货公司铁宫大楼。后来，这两个地方成为展示其作品的理想之所。他的设计中，标志性的 A 字裙让人想起墨西哥的特华纳风格。他也用墨西哥命名了他的不少设计，在这些设计中，我们能够看到来自当地传统服饰的波点印花、花饰和耀眼的金色刺绣。

1951 年，迪奥先生在威尼斯“世纪舞会”上将法国社会名流、美国《时尚芭莎》杂志巴黎版编辑黛丝・法罗斯装扮成了一位大胆的非洲女王，穿着醒目的豹纹和羽毛头饰。显然，非洲和古埃及是他的另一处旅行灵感所在。在一些迪奥先生创作的服饰上，你看得到克利奥帕特拉和拉美西斯名字的缩影。

似乎是沿袭着迪奥先生的传统，他的继任者们都是世界文化和旅行的爱好者。

虽然迪奥先生的门徒圣・洛朗并没有在迪奥时装屋执掌太长时间，但后来，他在自己的同名品牌中展示了其对中国、秘鲁、摩洛哥以及更广泛的异域文化的热爱。马克・博昂是这座时装屋里任职时间最长的设计师。在 1962 秋冬“箭形”高级订制系列中，他将“Air France（法国航空）”“巴黎东京”“曼谷之夜”和“里约除夕”等世界各地的灵感放在一起，以纪念旅行——这一时装屋的伟大传统。后来，在 1965 年春夏高级订制系列中，他将亚洲国家的文化重新诠释为丰富多彩的印花和长袍。非洲启发了博昂创造了 1964 秋冬包含了筒裙的高级订制和 1967 春夏“非洲风格”裙装和“猎装”套装，这位设计师没有采取老套的陈词滥调，而是将非洲的色彩和装饰抽象成非常简单和现代的轮廓，他的 1966 春夏高级订制系列也如法炮制。对华丽的热情被他倾注在 1968 年的两个系列中，设计师以大量华丽的刺绣、黄金饰品和长袍来重新书写拜占庭时代的辉煌。1962 年，博昂在孟买举办了该品牌的第一场时装秀，并设计了迪奥时装屋的第一件纱丽。27 年后，他的 1989 春夏“印度之年”高级订制“金字塔”及“纱丽”等系列以明亮的粉色、黄色和橙色为他在迪奥的任期画上了完美的句号。

意大利设计师吉安科罗·费雷曾多次在印度进行长途旅行。在迪奥时装屋的最后一年，他也将为自己的1996-1997秋冬高级订制系列向印度致敬，定名为“印度热情”。他以来自这个国度的色彩和珍宝为灵感，重新创造了刺绣、纱丽和头巾与一首浪漫色彩的史诗。费雷偶尔也会在他的设计中加入英国元素——这个国家也与时装屋的创始人有着不解之缘。1989-1990秋冬高级订制系列中，爱德华时代的上流社会和塞西尔·比顿的戏服设计为他的“皇家赛马会-塞西尔·比顿”系列提供了灵感；1990春夏高级订制系列中，莎士比亚的《仲夏夜之梦》在一整个轻盈如羽的系列中得到了完美诠释。当然，1992-1993秋冬高级订制系列之中，费雷以“威尼斯冬天的秘密”重访了他的祖国，让人们得以重新审视这座城市在18世纪的辉煌。

反叛的英国人约翰·加里亚诺可能是时装屋中最狂野的一位掌门人。他天马行空的想象力和旅行天衣无缝地融合在一起，借此创造了无数天桥上记入史册的造型和超现实的旅程。在他的世界里，旅行不受时间和空间的限制。非洲马赛部落会遇见欧洲的中国风情；上海的月份牌旗袍美女和好莱坞明星一道展示女性的魅力；一代妖姬玛切萨·卡萨蒂、玛塔·哈丽和蝴蝶夫人是他的旅伴；他与宝嘉康蒂公主和温莎公爵夫人一起乘坐迪奥的东方快车；还邀请奥地利的茜茜皇后穿越中欧到伊斯坦布尔旅行。横跨俄罗斯到蒙古，他不仅发现了部落风格的图案，而且还受到俄罗斯先锋艺术和中国红军的启发。葛饰北斋著名的神奈川冲浪图案曾装点过迪奥先生格兰维尔的别墅，加里亚诺则在他的2007春夏高级订制系列中，用布料折纸和一件令人惊叹的外套重现了这一艺术作品。极尽惊艳的2004春夏系列灵感来自于设计师在埃及的旅行，标志性的H形轮廓系列装饰着华丽的镀金色调，从纳芙蒂蒂到图坦卡蒙，法老和皇后们从帝王谷真正“起死回生”。

与加里亚诺相比，比利时设计师拉夫·西蒙似乎更加理性和冷静，但毫无疑问，和时装屋的创始人和他的前任一样，西蒙也是一位国际主义者。这位热衷现代主义的设计师更喜欢从更广泛的世界文化视角出发，而不是直接引用。通过重塑安迪·沃霍尔的艺术作品和邀请艺术家斯特林·鲁比加盟，他将美国艺术中最激动人心的部分注入了迪奥。在其2013-2014秋冬“世界高级时装”高级订制系列中，他展示了四个独特的文化灵感：“欧洲”部分着重于法国的传统，“美洲”部分着重于充满活力的大胆精神，“亚洲”展示的是“充满平衡、传统和纯洁的服装”，而“非洲”则落脚于“自由、玩乐和轻松的创造力”，同时也向马赛风格致敬。这些设计密码微妙而富含强大的力量。但只有当西蒙在2015年早秋时装秀上把我们带到东京的“大都市”时，我们才意识到他对日本文化的热爱有多深。卡哇伊风格和前卫的建筑灵感都在该系列中闪耀。他的旅行和想象涂抹着一层未来感的釉彩，幻想和现实共存。西蒙的目的地是——未来。

当玛丽亚·嘉茜娅·蔻丽来到迪奥时，她带来了她的意大利浪漫主义，但更重要的是，她带来了女性气质和女性主义。这位历史上第一位掌管时装屋的女性也面临着一个挑战，那就是如何界定文化挪用和欣赏之间的界限。如今，迪奥这家巴黎品牌的早春、早秋，甚至是高级订制时装秀在全世界上演，而蔻丽则利用这些机会向不同的文化致敬，并以此激励全球的女性。

2018早春度假系列在卡拉巴萨斯的一个露天保护区进行展示，她将创始人的拉斯科洞窟收藏品与加利福尼亚的荒野联系起来，收藏品的灵感来自于在法国西南部发现的古代洞穴壁画。而系列中，从新派的牛仔帽到格子花呢和流苏，西部风情元素无处不在。之后，当她了解到墨西哥马术比赛最初只允许男子参加时，蔻丽被那些墨西哥女骑手所感动和启发，邀请了八名女骑手前来秀场表演，并将这种精神转化为华丽的墨西哥图案、拉丁美洲魔幻现实主义文学等灵感。在许多其他的系列中，蔻丽通过将迪奥的标志性设计融合现代化的方式向迪奥先生致敬。例如，1953年的“日本花园”系列在东京展出的2017春夏高级订制时装系列中被重新创作，樱花和花鸟图案得以再现，作为专门为日本女性奉上的新设计。她还在上海的高级订制大秀上向迪奥先生的幸运红色致敬。1953 年，阿尔伯特 · 德卡里斯的一幅插图描绘了迪奥先生的旅行，引导她创作了 2018 秋冬高级订制系列。正如迪奥先生这位极具前瞻性思维的时装设计师在他的自传中所说 :“一个完整的系列应该面向所有国家、所有类型的女性。”而蔻丽正在将其实现。

虽然迪奥先生已经离开了我们，但这家法国时装屋和他的继任者们永远不会停止他们的探索之旅。他们将不断改变欧洲的时尚景观，使其变得更加包容和摩登，并致力于致敬世界各地的人民和精彩文化。

传承要靠创新

撰稿：洪晃

迪奥先生英年早逝，1957 年 10 月突然在意大利去世，那时候迪奥正处于巅峰时刻，先生用十年创造出一个传说，从 1947 年“新风貌”面世后，迪奥品牌就是女性美的代名词，奠定了一个世纪的时尚审美标准。

今天，迪奥先生已经离开我们 63 年了，然而迪奥品牌仍然继续领军时尚行业的审美标准，传说还在继续。

这不仅是迪奥先生的伟大，同时也是继迪奥先生之后，历届迪奥品牌艺术总监对迪奥品牌的诠释、对迪奥先生所建立的密码的分解和重构，当然最重要的是这些迪奥先生后人把自己的创意和时代的梦想也渗入了 1957 年来每一季的服装中。

讲述这些迪奥先生的传人和他们的设计也是“克里斯汀・迪奥，梦之设计师”展览很重要的一部分。

传承在中国往往是临摹或者复刻，一本《芥子园画谱》至今是中国国画的教材，学习方式就是临摹，越像越好。在我们的传统中，在前辈的作品上添油加醋，甚至篡改，都是大逆不道的。这便是“克里斯汀・迪奥，梦之设计师”展览在中国展出的重要意义：通过这个展览，我们清晰地看到迪奥品牌的传承不仅靠传承，同时必须创新。展览中详细介绍了迪奥先生之后六位设计师的作品和他们对迪奥品牌的诠释以及创新。这六位设计大师——圣・洛朗（1958-1960）；马克・博昂（1961-1989）；吉安科罗・费雷（1989-1996）；约翰・加里亚诺（1996-2011）；拉夫・西蒙（2012-2015）；玛丽亚・嘉茜娅・蔻丽（2016 至今），这六十几年来他们不仅继承了迪奥先生的品牌理念和 DNA，同时用自己的创意和对时代的理解，发扬光大了迪奥品牌的传说。“克里斯汀・迪奥，梦之设计师”展览是每个有品牌梦的中国人最好的教材，而对每一个参观者，这都是一场难忘的创意梦境之旅。

建设型大师

为了让迪奥的印章成为审美的标志，并得到几代人的认可和追捧，马克·博昂可谓立下汗马之功。

博昂自己说过："我做的一切都是符合迪奥传统的。"但是他也说过："我不知道如何去做迪奥先生，我非常敬仰他，但是我从未与他一起工作过，所以我从来没有复制他的奢望。我努力巩固迪奥先生最有代表性的元素，就是他作品的优雅和独特性的魅力，但是用我的风格去诠释。"这个听上去很矛盾的两段话精确地总结了马克·博昂在迪奥近 30 年的成功之道。就比如"Miss Dior（迪奥小姐）"这条裙子，迪奥先生在世的时候，几乎每季都会有一条以他妹妹命名的裙子。马克·博昂的"Miss Dior（迪奥小姐）"裙子保持了原轮廓，但是他在版式上都比迪奥先生更加宽松，加强了舒适感。马克·博昂的"Miss Dior（迪奥小姐）"裙，制造了 20 世纪最大的"撞衫"事件，1961 年两个巨星居然同款登台莫斯科电影节。这张照片可谓这次展览的一个看点。

"Gamin（顽童）"套装是马克·博昂修长风尚系列的经典，他的设计特点之一就是修饰女性身材，用今天的话说——显瘦。这些让迪奥品牌能够被更多女性享用，也为未来的迪奥品牌高级成衣奠定了基础。随着时尚行业的变迁，博昂于 1967 年开辟了迪奥高级成衣，乃至今天我们能够在迪奥店中挑衣服都是因为他当年的创举。

博昂的接班人吉安科罗·费雷刚开始不被看好，大概因为他是第一个要操刀法国第一时尚品牌的外国人。费雷是做成衣出身的设计师，但是对高级订制情有独钟。在去迪奥品牌之前，他自己在罗马曾经尝试做自己品牌的高级订制，而一旦去了迪奥，他马上关掉了自己的高级订制品牌，将全部精力投入迪奥品牌的建设。费雷与迪奥先生有两个共同的兴趣，一是对优雅奢华的表达，二是对建筑的兴趣。20 世纪 80 年代金融行业的兴盛带来了西方时尚对巴洛克风格的追随，而费罗将建筑结构元素和巴洛克风格融入夜礼服，展现出前所未有的优雅奢华。展览中费雷以罗马柱为灵感的夜礼服是最好的范例。然而费雷的工作风格和迪奥先生却绝然不同，他的到来给迪奥高级订制工坊带来新的工作方式，更加开放，更加民主。他喜欢和高级工坊的版师一起讨论服装，他甚至把工坊的布置全部改变了，更加开放，便于每道工艺之间的沟通，所以改变高级订制工坊的工作作风也是费雷在迪奥工作时的一大创新。

革命型的设计师

圣·洛朗和约翰·加利亚诺都属于革命型的设计师，两个人一个外向，一个内向，但是都有叛逆的创新风格，那种语不惊人死不休的创作风格。其实迪奥先生又何尝不是一个时尚的革命者呢！创新一直是迪奥品牌基因的一部分，曾几何时，巴黎女人在街上穿迪奥会被人撕的，真的，二战结束后，巴黎传统势力认为迪奥先生的设计是一种面料的浪费，当时年轻姑娘穿着迪奥的裙子上街，真的遭遇过被老女人手撕。展览中有图为证。因为那些二战刚过来的人看不惯年轻人对美的追求，认为是一种浪费。而迪奥第一季所展现的颠覆性的审美成为常规，就会有设计师要去颠覆迪奥的审美，其中包括迪奥自己的设计师。

每个浪子心中都有一份乡愁，而对于这圣·洛朗和加利亚诺，他们虽然叛逆，但是对迪奥先生的崇敬之心也是非常明显的，甚至感情化的。在宣布接手迪奥品牌艺术总监的发布会上，圣·洛朗说："我无法跟你形容我的所有感受，会用太长时间悲伤、苦恼，同时还有喜悦、自豪，还有怕自己失败的恐惧。"而当加利亚诺接手迪奥品牌的时候，说的话像孩子一般天真，他说："我就是想做迪奥先生会喜欢的时尚。"

但是两位设计师都曾经做出了迪奥品牌历史上最有争议的时尚，圣·洛朗倾情于20世纪50年代末的美国“垮掉一代”的文化，他在1960年将摩托车手的皮夹克纳入迪奥的时装秀，今天大家看到这件名为“芝加哥”的皮夹克，都会觉得非常迪奥，优雅、大手笔。但是当时的主流文化还是二战后的小布尔乔亚式审美，一件皮夹克在高级订制中出现绝对是会引发争议的。可惜圣·洛朗只做了六季之后就被征兵了，但是在这短暂的两年中，他让迪奥成为年轻人青睐的品牌。

约翰·加里亚诺在迪奥品牌很长时间，创举无数，而最受争议的一定是他以流浪者为灵感的一季，而也有传说，这一季的衣服是服装史上被拍摄最多的衣服。加利亚诺是服装史上最有戏剧感的设计师，可以说是最典型的明星设计师。他对服装的戏剧感很可能来源于他在1980年代初，曾经在伦敦大剧院当衣工。几年泡在剧院后台，使他清楚地认识到服装不仅是衣服，而是一种个性表达的语言，他可以用服装去评论一切。在他的设计中，既有布尔乔亚的优雅，但又暗藏着对这种优雅的讽刺。能够把时尚玩得反手如云覆手如雨的加利亚诺，真是非天才莫属。

除此以外他对剪裁有独特的技能，善于斜裁纺织面料，比如雪纺。纺织面料斜裁后廓形能力加强，1996年他为黛安娜王妃参加大都会时尚晚宴设计了一条斜裁的夜礼服，顿时惊艳。斜裁让梭织面料增加弹性，垂感更加贴身、性感。加利亚诺在迪奥的年代，迪奥品牌是众星追随的品牌，特别是加利亚诺的高级订制服装。每次迪奥秀场上的衣服都会在之后的一年中形成一场时尚风波。

哲学家设计师

当加里亚诺离开迪奥，明星设计师的时代也就结束了。接班的拉夫·西蒙和现在的玛丽亚·嘉茜娅·蔻丽都是知识份子型的设计师。两个人在做第一场秀之前都曾经花大量的时间在迪奥档案馆里面，不同的采访中，两个人都谈到迪奥档案馆就是一个宝库，一个灵感发源地。西蒙和迪奥先生对艺术有同样的激情，两个人都是在艺术中找到自己的内心，也都善于提炼艺术元素用于时尚。而蔻丽和迪奥先生一样对建筑独有嗜好,两个人都擅长用建筑结构方式去设计服装。21世纪，也是新型时尚面料层出不穷的时候，西蒙和蔻丽都很擅长用新的面料去诠释经典的迪奥元素。拉夫·西蒙的第一场高级订制秀，是向迪奥先生的第一场秀致敬的，所有的迪奥符号都扑面而来，但是又充满了惊喜，比如“花”，并不是在服装上，而是整个秀场五个房间的墙壁和天花板上铺满了鲜花，以至于秀还没开始，就已经被花海包围，空气中透发着花香。而西蒙的第一套衣服，足以把迪奥先生的“新风貌”在21世纪激活，这是一个黑色裤子套装，上身是迪奥先生经典的“Bar (迪奥套装)”，而下面却是一条裤子。满满的女性力量示意着一个新时代的迪奥女人。

蔻丽作为迪奥第一名女性艺术总监，更加扎实地把当代的女性主义思想和全球化视野带到了迪奥。蔻丽深信高级订制不只是为女性做一条漂亮的夜礼服，而是要赋予女性通过服装自我表达的能力。对于蔻丽来说，时尚必须要赋予女性力量这才是21世纪女性需要的。

蔻丽在她的设计中多次将“我们都是女性主义者”和“团结的姐妹有力量”这些当代女性主义的社会精神带入她的服装。她的科班出身和丰富的时尚经验，也造就了一位技术上非常成熟的艺术总监。蔻丽看似奢华的服装都有轻巧的结构，让奢华不再那么沉重。同时，她的日装抓住当代各种女性的特征，她所设计的服装是未来女装的方向，在继承迪奥品牌DNA的同时，用她独特的手段让华丽更加舒适，让时尚成为女性力量的源泉。

“克里斯汀·迪奥，梦之设计师”展览讲了很多故事，但是对于一个期望中国时尚品牌崛起的媒体人，这个展览有独特的意义，其叙述不仅是一个传奇的品牌故事，但是更多的是在传承中创新，在创新时发扬弘大一个与时俱进的传统。对于我来说这个展览是品牌传承最华丽的课堂，也是一个动听的传说。迪奥先生的丰功伟绩在他之后的六位设计师手中不断成为定义当下时尚审美的经典，不断创造传奇，而迪奥先生的故事也就成为最美丽的现代传说。也许，迪奥先生是对的，那颗幸运之星一直照耀着蒙田大道 30 号。

仙境晚礼服

撰稿：菊丝婷・皮卡迪

从克里斯汀・迪奥于 1947 年 2 月推出首个时装系列的那一刻起，浪漫主义就一直是他创作美学的核心，而精美的舞会礼服和晚装就是这种风格的绝佳代表。这些飘逸的裙裾中散发出的女性气质令人想起他的母亲玛德莱娜夫人在他童年时期所穿过的服饰，那是第一次世界大战爆发前和平繁荣的黄金时期，史称美好年代。当他在 20 世纪 30 年代后期开始设计时装的时候，另一场灾难性的战争已经逼近。迪奥先生于 20 世纪 40 年代初期在巴黎的吕西安・勒龙的高级时装公司工作，因此他很清楚德国占领会使法国时尚工业沦落为明日黄花。

在 1956 年出版的回忆录中，迪奥先生写道："由于战争期间穿着军装制服的影响，现在的女性看上去都和女武士打扮得差不多。我却为所有花样女子设计服饰，圆润的肩部轮廓，丰腴的上半身，只手可握的纤纤细腰配上巨大的裙摆。"[1] 如此优雅的轮廓需要使用大量的面料，被时尚界颇具影响力的《时尚芭莎》总编卡梅尔・斯诺称作"新风貌"。当然，要把这种设计放到战时纺织品配给制和战后经济紧缩的大背景下观看才能发现其中的革命性。迪奥首个时装系列中最受欢迎的晚礼服之一"Soirée（夜）"标志着好莱坞时尚的回归，采用海军蓝丝绸塔夫绸多层礼服裙，配低胸抹胸和缎带肩带。电影明星丽塔・海沃斯选择它出席电影《吉尔达》的巴黎首映式就是水到渠成了，影片中她饰演的就是经典的蛇蝎美人这类角色，迪奥先生的高雅创作风格使她的美国影迷们无不为之折腰。

迪奥先生随后的每次时装秀都向公众展示了同样充满魅力的设计，比如 1948 年春夏高级订制系列中"Fete（节日）"这条精致的粉红色拖地长裙，带有百褶裙撑；下一季的白色缎面舞会礼服"Virginie（薇吉尼）"带有天蓝色的罗缎腰带。随后的系列继续欢欣地歌颂美好，1949 年春夏高级订制系列中的"Miss Dior（迪奥小姐）"晚礼服就是裁缝手中诞生的迷人花束，裙子上盛开着一千朵绣花。

不过，克里斯汀・迪奥的风格从理念上并不仅限于青春洋溢的少女。如果说“Miss Dior（迪奥小姐)”晚礼服是受到他心爱的妹妹卡特琳娜的启发，那么1949年秋冬高级订制系列中充满戏剧性色彩的黑色天鹅绒晚礼服“Cygne Noir（黑天鹅)”则重现了他的缪斯米萨・布里卡尔所体现的成熟精致。这个系列堪称艺术大师的杰作，以其优雅而大胆的设计名垂青史，“Venus（维纳斯)”礼服淡灰色钉珠薄纱气韵雅致，与迪奥先生的杰作，大气的午夜蓝带亮片的“Junon（朱诺)”比肩而立。这些裙装已经超出了凡间女子的境界，而是为女神而设计的。

所以，玛莲娜・迪特里茜表示只穿着迪奥品牌的裙装，便可以理解了。她向导演希区柯克提出了条件“没有迪奥，就没有迪特里茜！”后才接受出演1950年的电影《欲海惊魂》中的女主角。她不仅是迪奥时装秀和时装大师乡村别墅的常客，也能够轻松驾驭大师的“Cygne Noir（黑天鹅)”等各种戏剧性设计。

迪奥品牌与世界各国王室的联系同样十分紧密。1949年春，年轻的玛格丽特公主首次出访欧洲大陆，途径巴黎时，她在蒙田大道的沙龙里与设计师见面，并订购了几件礼服。第二年，伦敦的法国驻英国大使馆举办迪奥小规模时装秀，她与母后和其他几位王室成员也都前来观看。1951年，玛格丽特公主为自己21岁生日挑选了一件童话风满满的白色迪奥舞会礼服，欧根纱裙和抹胸上都带有金色和珍珠母贝的刺绣，她宣布这是“自己最喜欢的裙装”，摄影师塞西尔・比顿为她这个形象按下快门，将此正式肖像载入史册。

克里斯汀・迪奥作为王公贵族和好莱坞明星首选的时装设计大师的地位也得到了格蕾丝・凯利的认可。她先是通过扮演电影角色实现女孩的美国梦，身着迪奥礼服走上红毯，形象完美，后来她又嫁入豪门成为王妃，1956年1月在与摩纳哥大公兰尼埃的订婚典礼上，她身穿白色缎面迪奥礼服，形象优雅至极，成为一代经典，她一生都是迪奥品牌的忠实顾客。

克里斯汀・迪奥认为：“晚礼服是梦想，它也必须让你成为他人的梦想。”他创造幻想和愉悦的天赋在1951年9月在威尼斯拉比亚宫“世纪舞会”上达到了新的高度。东道主是富有的社交达人贝斯特古，他于1948年购入这座宫殿后花了近一百万美元进行维护装修，使它成为1500位宾客尽情展示自己的舞台，《生活》杂志将来宾定义为“世界上最高贵的蓝血统且最富有的居民”[2]。克里斯汀・迪奥是应邀参加舞会的宾客，他还为许多宾客设计舞会礼服，《时尚芭莎》当时驻巴黎记者、嘴尖牙利的富有名媛、红毯杀手黛丝・法罗斯就是其中一位，据参加舞会的戴维・赫伯特回忆，“黛丝的礼服是用好几码长的香槟色雪纺制成，还隐隐印着豹纹呢”。至于迪奥先生本人，他在回忆录中这样写道，“贝斯特古舞会是我从未见过以后也不会再见到的最神妙的奇观。宾客身上的华服比墙上提埃坡罗壁画中人物的奢华服饰毫不逊色。意大利某个仲夏夜的魔力使我们陷入永恒的咒语，帮我们摆脱时间的约束。”[3]

由此，迪奥现任艺术总监玛丽亚・嘉茜娅・蔻丽为2019年5月在威尼斯拉比亚宫假面舞会制作礼服也便顺理成章，卡莉・克洛斯和西耶娜・米勒等宾客穿着她设计的高雅而飘逸的礼服裙翩翩起舞。那天晚上，玛丽亚・嘉茜娅为迪奥品牌续写传奇，她凭着直觉感到，一件晚礼服可以给人带来这么多幸福和欣喜。克里斯汀・迪奥在回忆录的末尾写道：“当今世界上，高级订制时装是最后一个存放仙女魔法的宝库，而时装设计师则是灰姑娘仙女教母的魔杖的最后一位使用者。”值得庆幸的是，迪奥的魔杖经过多年的传承，至今仍像它七十多年前首次发挥魔力时那样神奇。

1 克里斯汀・迪奥：《迪奥自传》，维多利亚和阿尔伯特博物馆出版社，2012年：22–23页。

2 《生活杂志见证威尼斯某宫殿开幕：墨西哥百万富翁举办年度最大派对》，载《生活》杂志1951年9月24日号：173页。

3 克里斯汀・迪奥：《迪奥自传》，维多利亚和阿尔伯特博物馆出版社，2012年：36页。

4 同上：193页。

CONTENTS

INTRODUCTION: CHRISTIAN DIOR'S NEW DREAM LOOK

ORIOLE CULLEN

The exhibition *Christian Dior Designer of Dreams* charts the arc of the House of Dior, from its inauguration in 1946 through to its position as one of the world's leading fashion houses today. Through historic garments, documents and imagery from the archives of Dior Héritage and objects from museums, galleries and private collections around the world, the exhibition brings to life the intriguing and multifaceted history of the House. Illuminating the life and work of the founder, Christian Dior himself, and celebrating the creative directors who have come in his wake, the exhibition and this accompanying publication of selected essays champion the essence of haute couture developed and created over seven decades by the House of Dior.

The story of the House of Christian Dior begins in Paris at number 30 Avenue Montaigne, an elegant Parisian townhouse building which today remains the home of Dior. In his autobiography, Christian Dior – who had considered a career in architecture as a young man – claimed he had long been fascinated by this *hôtel particulier* with its "neat, compact proportions, and air of sober elegance without the least hint of ostentation".[1] When textile magnate Marcel Boussac approached Dior and offered to finance a couture house in his name, the designer was able to obtain the lease to his much-admired building. He revelled in the project of renovating the house, employing his friend, interior decorator Victor Grandpierre and working together to decorate it in colours that would become so associated with the House of Dior "all in white and pearl grey, looking very Parisian".[2] It was to this building, on 12 February 1947, that crowds of fashion editors, buyers and potential clients would flock to witness a key moment in the history of fashion: the launch of Christian Dior and the collection which would become known as the "New Look".

As Justine Picardie points out in her essay on the romance of Dior's eveningwear (pages 65–67), the impact of the "New Look" was so strong because it came directly from a period of hardship and austerity during the Second World War years of 1939–45. Fashion was one of the most important industries in France at the start of the war, with some 300,000 people employed in haute couture and its ancillary trades. Under the Occupation many of the couture houses shuttered and Lucien Lelong, head of the Chambre Syndicale de la Haute Couture, battled to foil a plan by the German authorities to move the industry to Berlin. When the war ended, the fashion trade was in a much-reduced state having lost the patronage of overseas buyers, particularly those from the all-important American market. By the time Christian Dior set up his couture house in late 1946, the fashion world was watching Paris, waiting to see if it could regain its crown as the centre of style.

In this atmosphere, there was much expectation and comment in the press in the lead-up to Dior's first show. He was already known to the fashion press as a designer at the house of Lelong where he had previously worked with fellow designer Pierre Balmain. The excitement had been building since the summer of 1946 when the first news of Dior's new business had leaked. The influential editor of American *Harper's Bazaar*, Carmel Snow was a great fan of Dior's work and along with American *Vogue* editor Bettina Ballard, a close friend of the designer, she had promoted Dior in advance, provoking great anticipation amongst the American press and buyers.

On the day of the show the crowds battling for entry to 30 Avenue Montaigne demonstrated the frenzied interest in this new house that had been stirred up. Dior's presentation was a revelation. The rooms had been scented with what would become Dior's first perfume, *Miss Dior* – a story discussed in depth by Catherine Zeitoun (pages 45–48) – and the audience was amazed by the confident way in which Dior's models strode through the rooms of the house. *Harper's Bazaar* editor Anne Scott James recalled the models were "brushing the audience carelessly with their skirts… walking rapidly, smiling at no-one and stopping for no one. As the great skirts swirled, the pleats flew, the taffetas rustled, the embroideries sparkled and shone, the colours mingled and fused."[3] These luxurious garments and the two new silhouettes *Corolle* and *En 8* introduced by Dior truly captured the excitement and delivered on the anticipation of all present. *Corolle* took the shape of an upside-down flower with the bodice featuring a narrow, sloping shoulder and tiny waist and the hips padded above a full flaring skirt which fell far below the length of the current fashions. Some of these skirts were made from over 13 metres of fabric, a shocking display of luxury to an audience who were still living through the lingering austerity of the war. *En 8* differed in that the narrow skirt, hugged tightly to the legs and with a tiny waist, created a silhouette resembling the number 8. The fashion world was thrilled with this collection. Carmel Snow crowned it the "New Look".

As Jerry Stafford describes in detail (pages 49–54), many of the attendees were taken with Dior's subtle colour palette. British fashion journalist Winifred Lewis wrote of Dior's "particular genius for colour"[4] and numerous magazines detailed his unusual and unique combinations and shades. *Elle* magazine wrote of his use of blue-greys, ash and slate, like the sky reflected in a puddle on the ground, and grey-beiges echoing the ash-laden paths of the Tuileries gardens.[5] Yet, it was a black-and-white afternoon suit which would become the emblem of the New Look, its graphic appearance perfect for reproduction in newspapers and magazines at a time when printing colour images was still the exception. Titled *Bar*, this suit, created by the young Pierre Cardin, then head of Dior's tailoring workshop, consisted of an off-white tussore silk jacket with sloping shoulders, a narrow waist, and hips that were carefully padded to spring outwards over the skirt. The skirt itself was full and long and made from finely pleated black wool. The suit was crowned with a large black picture hat. *L'Officiel* included an illustration of the suit in its report on the Spring collections, and British *Vogue* published a drawing of the suit by Dior's close friend Christian Bérard, remarking on the "tight tussore jacket padded to a tea-cup curve".[6] In America, *Harper's Bazaar* reproduced an image of the *Bar* suit by another great friend of Dior's, illustrator René Gruau, stating "Christian Dior's Paris afternoon suit, 'Bar' holds the key to the new fashion – the bell-shape, exaggerated hipline in the natural tussore jacket, the longer pleated black wool skirt."[7] A full-page photograph by Serge Balkin printed in American *Vogue* showed a model wearing the *Bar* suit standing sideways on the staircase of 30 Avenue Montaigne. The *Bar* had been well and truly chosen by the fashion world to represent this New Look.

When in 1955, eight years after the launch of his House, Christian Dior gave a lecture at Paris's La Sorbonne University accompanied by models wearing his creations, he made sure to include the *Bar* suit in the line-up. The suit was captured that same year in a photograph by Willy Maywald, worn by model Renée Breton, the original black hat replaced by a natural-coloured straw hat. This photograph would go on to become the definitive image of the suit. For every one of the six subsequent creative directors who have led the House of Dior, since the untimely death of the founder in 1957, the *Bar* suit has remained an important emblem and one which is continually referenced in many different iterations. Huang Hung (pages 61–64) looks at this legacy and touches on how each director offers a constant reinvention of the Dior signature.

The runaway success of the New Look and the ushering in of a new era of femininity is much debated. There can be no doubt that Dior loved women. He was very close to his sister Catherine and surrounded himself with and paid tribute to key female colleagues within his couture house. His right-hand woman was Raymonde Zehnacker whom he referred to as a "second-self" and as someone who held "the reins of the business in her firm and capable grasp."[8] Marguerite Carré held the crucial role of Technical Director of the ateliers – the very heart of the house, as discussed by Nadia Albertini (pages 39–43) – where the signature look and cut of Dior was created as garments were physically brought to life. The third person in this important triumvirate was Dior's muse Mizza Bricard who inspired his creations and oversaw his millinery. Some, however, questioned his silhouette with its narrow waist and long skirts as reverting to an earlier time when women were more restricted. Now they were used to the freedom of the war years where trousers, shorter practical skirts and uniforms were popular forms of dress. The magazine *Britannia and Eve* commented on this new "war" between those who loved the new fashions and those who protested, noting that all of the publicity about the issue "is probably achieving more for the cause of the longer skirt and the pinched-in waist than anything else. It has made women aware of them and awareness of a new fashion is only one step to wearing it."[9] And women wanted to wear the New Look. Reflecting on her long career in fashion, the journalist and friend of Christian Dior, Ernestine Carter noted that for her "Nothing has ever come up to the exhilaration of those first Dior collections, and never has so universally becoming a fashion nor one so enduring been devised. Tall women, short women, large women and small women, older women, young women the New Look suited them all."[10]

Today, this idea of encompassing and appealing to all women is very much at the heart of Dior's Creative Director Maria Grazia Chiuri. The first woman to helm the House of Dior, she has demonstrated the possibility of combining both feminism and femininity with collections inspired by feminist writers, artists and activists who inform ideas around her work. The exhibition displays her own version of the *Bar* suit, with a feather-light transparent tulle skirt and a "Bar" jacket, open to reveal a T-shirt printed with the words of Chimamanda Ngozi Adichie: "We should all be feminists". Introducing an element of lightness and modernity through the context of Christian Dior's original designs, Chiuri manages to stay true to his ideals and silhouettes whilst radically translating Dior's "New Look" for a new generation.

[1] Christian Dior, *Dior by Dior* (London: V&A Publishing, 2012), p.10.
[2] *Ibid.*, p.21
[3] Anne Scott-James, *In the Mink* (New York: E.P. Dutton & Co., 1952), p.120.
[4] Winifred Lewis, *The Tatler and Bystander*, 26 March 1947, p.382.
[5] *Elle*, 4 March 1947, p.9.
[6] British *Vogue*, April 1947, p.50.
[7] *Harper's Bazaar*, May 1947, p.130.
[8] Christian Dior, *op. cit.*, p.12.
[9] Winifred Lewis in *Britannia and Eve*.
[10] Ernestine Carter, *With Tongue in Chic* (London: Michael Joseph, 1974), p.76.

THE ART OF THE ATELIERS

NADIA ALBERTINI

Since the 1980s, the House of Dior has set up the Dior Héritage archives, an inexhaustible source of inspiration for its creative directors, who regularly peruse these treasures. The entire DNA of the brand is found there, preserved in the hundreds of silhouettes, the lines, the materials and the colours. Finding inspiration in designs is an opportunity to learn more about the history of the firm and the craftsmanship of haute couture.

For the Spring-Summer 1949 collection, known as *Trompe l'œil*, Dior designed an evening dress called *Miss Dior*, a nod to his first fragrance which was launched in 1947. The notes of the scent – a mixture of elegant chypre with innovative green accents – became petals and pistils, created by the Judith Barbier workshop. In *Haute couture, terre inconnue*,[1] Célia Bertin investigates the profession and introduces us to the work of this artificial flower maker, one of the last still active in Paris in the 1950s. The workshop at 18 Rue Daunou had been set up in 1845 and retained its founder's name. Mentioned as early as 1916 in the periodical *Les Elégances parisiennes*, it produced all kinds of decorative items made from artificial flowers. Embroidery, headdresses, garlands and brooches are all mentioned in a promotional lithograph that was published in the *Annuaire du luxe à Paris*, edited by Paul Poiret, in 1928. The studio offered more than six thousand different flower styles to the couturiers every year. Their working method had remained unchanged since the turn of the century; the workers, all women, still starched the fabrics by stretching them on enormous wooden frames. Once dry, the fabrics were folded, then cut up using cutting dies and a huge mechanical press. There were many different shapes: individual petals, whole flowers, leaves, branches – all that was needed to please the designers.

For *Miss Dior*, the *petites mains* embroidered test samples, covering the fabric completely in perfect imitations of brimming flowerbeds full of blooms. On their worktables they would arrange dozens of petals in a variety of shapes, sizes and colours, and would bring these tiny morsels of fabric to life by pressing them in veining moulds that dated from the 19th century. They might also form the petals one by one, by hand. Shaping a flower consisted of using a flame to heat a spherical metal tool that could be a variety of sizes. The tool was then pressed onto the fabric, which had been placed beforehand on a soft, slightly damp pad. This procedure gave each petal a form that was more or less rounded, depending on the size of the tool and the pressure that was applied. In this embroidered assortment, the various elements of *Miss Dior* can be found: powder-pink carnations with uneven edges, delicate jasmine flowers and orange blossoms. Daisies, violets and sprigs of lily-of-the-valley – Dior's lucky flower – complete the bountiful bouquet. Set side by side, these individual beauties form a perfectly harmonious ensemble.

In 1949, the Judith Barbier workshop was striving to imitate true nature – the nature that fascinated Dior. This traditionally Parisian know-how has been handed down and survives to this day in the Lemarié firm. Founded in 1880, the workshop is now the sole custodian of this hand-memory. Seventy years later, the mission that is entrusted to them is slightly different. Yes, it is still a matter of mimicking. But this time through taming a nature on which the passing years have left a particular patina. The dyes used on the archive piece have slightly faded, and some colours have oxidised. So the flower makers set to work: they closely examine the original dress, which has been specially lent to them by Dior Héritage for the occasion, from every possible angle. They identify and note the types of flowers, and the sizes of the pistils and seeds. They prepare, cut, shape and fix, petal by petal. Then, guided by the sound advice and watchful eye of Mr Liets, Director of Embroidery at Christian Dior, they dab the embroidered panels with a tea-based dye. For a better final effect, they slightly crush the result of endless hours of work.

Haute couture is certainly an art, but it is first and foremost a story of men and women, of interaction and collaboration. The renowned Parisian artisans make Christian Dior's vision a reality. Barbier might be mentioned here, but also Ginisty, Lanel, Lemarié, Lesage, Métral and Vermont. The couturier professed to greatly respecting this craftsmanship that enriched his creations. Their traditional savoir-faire and the rarity of the materials served only to heighten the sumptuousness of his dresses. Reading what he said in the speech he gave to over 4,000 people at the Grand Amphitheatre of the Sorbonne on 3 August 1955, the close connection he had with these flower makers, feather workers and embroiderers is plain to see:

> The desire to apply ornament without purpose is not of our time: the infinitely varied range of ornamentation – buttons, embroidery, ribbons, passementerie – allows us to specify the meaning, the spirit that we want to give to our dresses. In the past, heraldic devices would be embroidered on doublets just as they were sculpted on breastplates. While ornament may have lost its symbolic value for us, it remains an integral part of the dress, and is not a mere "add-on". And since Fashion is made entirely of contrasts, the ornament, through its material and its more or less enduring nature, contrasts with the dress that it completes. Thus a thick woollen suit will be fastened with fragile buttons, and a material as light as tulle will be studded with high-relief embroidery in gold or silver thread or with pearls.

It was a piece of embroidery by René Bégué, known as "Rébé", that inspired Maria Grazia Chiuri for her first Haute Couture collection. During a visit to the Dior Museum in Granville, Normandy, the designer spotted an embroidery sample that was dated 1954. Rébé, the embroidery workshop directed by René Bégué and his wife Andrée, was one of the greatest embroidery firms. For more than 60 years, they dreamt up sumptuous designs for the top couturiers. They were great friends of Christian Dior's, sharing his passion for flowers, and they received more than 250 commissions from him between 1947 and 1957. Dior was won over by Andrée's amazing eye for colour and by René's experience. Through his sketches and his romantic vision of the *femme-fleur* (flower-like woman), he enabled them to fully express their creative talents. Rébé in turn inspired Dior with their hundreds of floral motifs and samples imbued with the spirit of the 18th century.

The chosen mock-up, gifted to the museum by Rébé's heirs in 1988, shows a female silhouette typical of the 1950s – ample bosom, cinched waist, "Corolle" skirt – adorned with springtime motifs. Chiuri was immediately captivated by it, and attributed it to one of the designs for the Spring-Summer 2017 collection. The Rébé style is characterised by a subtle mingling of traditional drawing in the design and a constant search for innovation that results in the embroidery perfectly integrating into the silhouette. The landscape in question is made of a mixture of multicoloured raffia, cotton and silk threads. The irises, gladioli and daisies depicted by Rébé are embroidered using about ten different stitches. This diversity gives unique texture and relief. The whole thing is punctuated with a subtle shimmer, like a fine layer of morning dew, conveyed by iridescent sequins hidden here and there amid the threads. This sample, with its extraordinary technical and stylistic richness, served as an inspiration for the Safrane Cortambert workshop in the making of the *Essence d'herbier* strapless dress, which is now preserved at Dior Héritage and also at the National Gallery of Victoria in Melbourne. The contemporary interpretation that Chiuri has conceived, a true tribute to luxury craftsmanship, is showcased in the film for the 2019 advertising campaign for the *Miss Dior* fragrance. Worn by Natalie Portman, we can admire the great modernity of this silhouette, embellished with this blend of wildflowers in motion.

The magic of the workshops does not stop with the suppliers. On the contrary, it resides within the House of Dior itself. Florence Chehet and Nadège Guenin, heads of the Flou and Tailleur atelier respectively, unveil the secrets of the "construction" of the designs. They underline the importance of "Dior words", such as "Bar line" and "lightness" – words that are loaded with identity and that are forever reverberating around this temple of couture.

Dressed in a plain white coat – the same as the one that Christian Dior used to wear when he was drawing in his studio – over a hundred people work in this enormous hive of activity. There are 60 employees in the Fine Soft Dressmaking atelier and 40 in the Tailoring one. All ages mingle together: from 58 years old for the most experienced worker, to 17 for the youngest apprentice. Each of them has the same tools, which are surprising in their simplicity. A little leather pouch, hand-made by each of them, is worn around the waist or around the neck. It is equipped with needles, pins, safety pins, a pair of scissors, a thimble and a tape measure. It is straightforward and efficient, as is everything in these ateliers that Chehet and Guenin direct so deftly. They have been working for Christian Dior since 1982 and master every fabric, every stitch, every fingertip technique with both eyes closed. Like fine artisan workshops, they adapt their knowledge and their teams' talents to translate each creative director's vision.

The technical principles of haute couture are still the same after all these years. From pattern-making to bias-cutting to hemming, everything is done by hand, except for a few pieces of straight-line stitching that are done with a machine. After the client has had a first fitting and her measurements have been taken, a mannequin is "installed". The team reshapes a Stockman mannequin in accordance with the client's body. Curves are added where needed, or it is slimmed down; it is adjusted to perfection and used as a basis with which to work on the commission concerned. When making haute-couture pieces, extreme care and infinite precision are brought to every stage. The pattern-maker produces a toile which, once approved, will become a pattern. The *Miss Dior* dress was composed of a boned bustier and a full skirt, in line with Monsieur Dior's wishes. Making a dress in two distinct parts allows greater comfort and more elegance when walking. The design is made of two layers of ivory tulle, an ultra-light and transparent material, perfectly diaphanous, under which are placed petticoats in pinkish-beige organza and tulle, giving the skirt its shape and grace. The bustier is lined with salmon-pink silk taffeta for enhanced comfort. Despite the extremely sophisticated assembly technique, all Dior dresses must offer a light, feminine look, as *Essence d'herbier* demonstrates. The "construction" of each of these two dresses alone represents 300 hours' work, without counting the time spent on embroidery. Attaining perfection demands time. Enough time for the eye and the hand to become one, and to develop the intuition of touch: such are the so jealously guarded secrets of haute couture.

During haute-couture catwalk shows, the effect is striking: the dresses appear as if by magic, born from a combination of virtuoso hands from the various workshops. The inextricable link that has existed between the House of Dior and the craft professions ever since 1947 rises to the challenge once again. The endlessly repeated gestures that fashion the material with so much talent are crucial to haute couture's radiance. This type of intelligence has an element of the sublime that Christian Dior managed to express in his memoires: "Haute-couture dresses (...) are among the last remaining things to be made by hand, by human hands whose value remains irreplaceable for they endow everything they create with qualities that a machine could never give them: poetry and life."[2]

[1] Célia Bertin, *Haute couture, terre inconnue* (Paris: Librairie Hachette, 1956).
[2] Christian Dior, "Comment on fait la mode ?", *Le Figaro Littéraire*, 8 June 1957.

THE SCENTED GARDENS OF CHRISTIAN DIOR

CATHERINE ZEITOUN

For Christian Dior, gardens were taken for granted, ever since his early childhood at his family's house in Granville, Normandy. He was fascinated by nature – its shapes, its colours, its aromas. He developed a passion for botany and spent hours in the garden around the house with his mother. "A passion for flowers inherited from my mother meant that I was at my happiest among plants and flower-beds."[1] Flowers would be a feature of his family relationships, first with his mother and then with his sister Catherine, as he shared his love of gardening with her in Callian, Provence. "I found myself living for the first time in the depths of the country. I became passionately fond of it and developed a feeling for hard labour on the land, the cycle of the seasons, and the perpetually renewed mystery of germination. (...) I had a strong streak of the peasant, so with my sister I decided to cultivate the little piece of land which surrounded the house."[2] The periods he spent in this region drove him to buy the Château de la Colle Noire, not far from there, in 1950. Connected to the land, sensitive to the cycle of the seasons and mad about gardening, for Dior the countryside was a synonym with rest, and he looked forward to these times to take a breath between each collection. These moments beyond time, away from all thoughts of fashion and far from the city, allowed him to find new sources of inspiration, starting with his very first collection.

He founded the Parfums Christian Dior perfume company from the very outset of the fashion house, in 1947, in collaboration with his childhood friend Serge Heftler-Louiche, who had previously managed the Coty perfume and cosmetics firm and was thus already well versed in the perfume world at Coty. They wanted a Dior olfactory signature in symbiosis with the "New Look" collection. "And then Miss Dior was born. It was born during those evenings in Provence flashing with fireflies, where green jasmine served as a counterpoint to the melody of the night and the earth."[3] Christian Dior and Mizza Bricard came up with the name together. The latter exclaimed one day, when Catherine Dior walked in unexpectedly: "Look, here's Miss Dior!" to which he replied: "Miss Dior! There's my perfume."

Miss Dior, "who must smell of love", was one of the first green chypres in the history of perfumery after *Le Chypre* by Coty and *Vent Vert* by Balmain. It was created by Paul Vacher. The aldehydic, aromatic top notes open out into a bouquet of rose, gardenia, iris, jasmine and narcissus before finishing on notes of chypre, amber and leather. Being considerably more concentrated than the average perfume, it is a profusion, a mass of sprays of fresh flowers – a reflection of the vast bouquets of blooms that adorn the salons on the day when the inaugural catwalk show takes place and of the miles of fabric used for the typical dresses of the collection's famous *Corolle* silhouette. Its original bottle, designed by Christian Dior, took the form of an amphora, a symbol of the container used in antiquity to preserve fragile and precious commodities. This shape thus repeated the other silhouette of the Spring-Summer 1947 collection, *En 8*. "Jealous of having the same stamp, the dress and the bottle link up and get together. The expensive vial is narrow-waisted, the dress is curvaceous."[4] The fragrance's packaging was dressed up in a houndstooth motif, and Christian Dior asked René Gruau to design the advertising. A limited edition launched for Christmas 1952 took the form of Bobby, Christian Dior's dog, and was designed by Fernand Guery-Colas, with a box in the form of a kiosk.

Diorissimo was created in 1956 by Edmond Roudnitska. This iconic floral fragrance is a symphony that borrows the language of music and of the superlative, as if to signify Dior's world taken to extremes. Born from the meeting of its two creators – Dior and Roudnitska – it is one of the perfumer's most accomplished creations, as in it he found his style and his stamp by pushing the process of refinement to the maximum. The liquid and the bottle are in harmony just as these two geniuses' tastes and savoir-faire were. This was the perfume that Christian Dior cherished most, the one that recreated the scent of his favourite flower, his lucky charm. He wore lily-of-the-valley as a buttonhole, carried it in a reliquary that he secretly placed in his jacket pocket, and borrowed its name, shape or motif for dresses, jewellery, hats, shoes and even an entire collection: that of Spring-Summer 1954. For Roudnitska, too, lily-of-the-valley was an obsession. It was planted in his garden at Cabris – near the French perfume capital, Grasse, in Provence – and he would smell it in the springtime, for many years, every day that it was in bloom, and would note its variations, its different aspects, and the impressions that it conjured up for him. It is the fragrance of luck, sweetness, freshness, spring's awakening. Far from being a single-flower scent, as one might think, it is a symphony of flowers in conversation, ylang-ylang in dialogue with orange blossom, jasmine and Grasse rose for an evocation of a muted flower that will never be hushed again. The bottle, again in the shape of an amphora, was designed by Christian Dior and produced by Baccarat. It is crowned with a gold-plated bronze stopper, around which curl stems of lily-of-the-valley. It was to be the last perfume that Dior would create in his lifetime. A fashion world in mourning covered his grave with thousands of lily-of-the-valley sprigs on 29 October 1957.

In 1966 *Eau Sauvage* was born, a product of Roudnitska's imagination: "The perfume that, through its discreetly floral but enduring freshness, perfectly symbolises youth."[5] Its simplicity and freshness marked out a new style of fragrances that would never go out of fashion: that of unisex scents that are very often citrusy like eau de Cologne, a type that had already been initiated with *Eau Fraîche* in 1953, which had stolen the crown from lavender water in Christian Dior's heart and on his skin. It was the first perfume in history to use hedione, a synthetic molecule that is naturally present in jasmine, bringing freshness and strength to a floral bouquet. *Eau Sauvage*'s "discreetly floral freshness" inspired Roudnitska, in 1972, to create one of his favourite perfumes: *Diorella*. It was the first light floral fragrance with fruity notes. After an introduction of Sicilian lemon, honeysuckle is then unveiled in a dialogue with peach, and jasmine follows in their wake, bright and clear as water.

Poison, created by Édouard Fléchier, was the olfactory shock of the 1980s. This floral scent is led by tuberose, tinged with rose and carnation, which gives way to notes of fruit (plum) and spice (star anise, cinnamon) to compose an indulgent floral fragrance of the sort that will always be in vogue.

As for the *Dune* fragrance, formulated by Jean-Louis Sieuzac in 1991, it evokes a sandy seaside landscape. Its spicy notes bring warmth to this floral perfume with its overtones of water and ocean breezes constructed around the scent of peony.

J'adore, created in 1999, is the work of Calice Becker. This perfume echoes Jean Cocteau's description of Christian Dior: "This gentle genius of our times, whose magical name carries those of God (*Dieu*) and gold (*or*)."[6] It associates a harmonious blend of flowers with "a basket of ripe fruit" and a note of prunes in Banyuls wine. *J'adore* reconnects with the idea of excess that is present in *Miss Dior*; this time it was no longer about infinite lengths of fabric, but a shower of gold at the close of a decade that was marked in fashion and perfume by a tendency to pare down. Christian Dior had given John Galliano the space to become a fashion history icon by putting his stamp of exuberance on the venerable firm and designing Baroque, theatrical fashion imbued with a thousand historical and ethnic influences. The *J'adore* bottle revisits the amphora form that Christian Dior so loved. It was designed by Hervé van der Straeten and sports a neck ornament with a tribal flavour, as used by John Galliano in his first collections. Later, it would lead to variations conceived by François Demachy and entitled *J'adore absolu*, *J'adore l'Or*, *J'adore in Joy* ...

Demachy, a perfumer for the firm since spring 2006, is the originator of the collection for Parfums Christian Dior that mingles historical fragrances with numerous new creations, always bringing out the best of the exceptional flowers from the Grasse area and the sumptuous raw materials. As the custodian of a magnificent olfactory heritage, he oversaw the continuity of the historical scents on which the firm was founded, and dreamt up new olfactory symphonies around the flowers that the Dior family so loved. The *La Colle Noire* perfume pays tribute to the cabbage rose so beloved by Christian and cultivated by his sister Catherine at Callian. As for *Lucky*, it follows in the footsteps of the lucky-charm fragrances that began with *Diorissimo*, while at the same time conjuring up Christian Dior's favourite flower: "It's the perfume for all those times when you want to cross your fingers."

Dior's perfumes past and present are the result of the inspiration that their creators have found in the perfume plants of the Grasse area, lovingly cultivated on land that is conducive to their growth and in sublime landscapes that Christian Dior, Catherine Dior and later François Demachy have all so adored. Today, these raw materials are protected by their farmers, who have got together to form an association to safeguard this know-how that is now inscribed on UNESCO'S List of Intangible Cultural Heritage so that these aromas of royal jasmine, cabbage rose, tuberosa, orange blossom, iris, narcissus and mimosa so precious to the Christian Dior perfumes can never disappear. Thus there has been a continuity between the different noses of this illustrious firm, which is fêted and acclaimed for its floral fragrances, from *Miss Dior* to *J'adore* – sometimes figurative, sometimes abstract, striking in their truth and clarity. Inspired by these much-loved flowers, as soon as they are released from the bottle they conjure up a world of scented gardens.

[1] Christian Dior, *Dior by Dior* (London: V&A Publishing, 2012), p.168.
[2] *Ibid*., p.186.
[3] From an interview with Christian Dior, circa 1950.
[4] Colette on couturiers' perfumes, in "Fragrance", preface to the advertising brochure *L'Opéra de l'odorat*, published for Lanvin perfumes, 1949.
[5] Edmond Roudnitska, quoted in *Christian Dior, l'autre lui-même*, 1987.
[6] Vincent Leret, "Christian Dior parfums, 70 ans de parfums", *Christian Dior Esprit de Parfums*, exhibition catalogue (Milan: Silvana Editoriale, May 2017), p.59.

COLOURAMA

JERRY STAFFORD

"I have no wish to deprive fashion (and the ladies) of the added allure and charm of colour." – Christian Dior

"Colour may be used in touches if you wish to change the way of your clothes." – Christian Dior

In 1947, the American Abstract Expressionist painter Jackson Pollock produced the first of his groundbreaking "drip paintings", blonde bombshell Marilyn Monroe made her silver-screen debut, and a huge meteor created an impact crater in the mountain range of Sikhote-Alin deep in the heart of the Soviet Union. That very same year, Christian Dior caused an equally seismic shift when he announced his own revolutionary manifesto of cut and colour, his New Look: the debut and defining collection in which he presented the garment as "a piece of ephemeral architecture, dedicated to the beauty of the female body".

All artists – whether designers, painters or composers – understand the power of colour, its energy and the almost mystical effect it has on the eye and soul. They understand that when we perceive any colour, what we are really seeing is a process of interaction. Colour is a language that contains information and has the power to communicate within a society as effectively as it does in the colour-saturated natural world.

From his very beginnings Dior displayed a sumptuous palette, a unique colourama of shades and tones, each with its own personal history and meaning for both the couturier himself and those who subsequently took up his mantle.

The designer's sensational debut collection, as with all the collections designed for Christian Dior to follow, is a passionate ode to colour in which he introduced his clientele to a sophisticated spectrum of hues, ranging from delicate whites, pastoral pinks and blossoming yellows, to sylvan greens, mysterious blues, silent greys and elegant blacks, punctuated with those signature red exclamations, his "*Trafalgars*", soaring like operatic arias or masterful strokes from a painter's brush.

His favourite "*chéries*" would parade before the hushed salon audience and a meticulously gloved hand would stifle a gasp of admiration, astonishment or even shock as a "*rouge éclatant*" (bright red) or a "*rouge satan*" (Satan red) electrified the first *passages* of this New Look (Autumn-Winter 1947); and as the models advanced through the salon, the blood-red ink of René Gruau's quill transformed their poses into unforgettable illustrations for fashion magazines around the world.

The iconic colour would become one of the designer's recurrent passions, and he attributed it with a particular luck and charm. Red, corresponding with fire, also symbolises good fortune and joy in Chinese culture.

Designer Maria Grazia Chiuri continues to respect this heritage today with her own significant approach to colour, extolling like Dior himself "precise and intentional colours which communicate an important meaning for those who choose to wear Dior today" while the pursuit of the perfect red continues for the Creative Director of cosmetics Peter Philips, for whom red will always be "the colour of passion, of love, of life and of Dior!"

Of course Christian Dior did not only experiment with his palette on the catwalk; he also introduced a groundbreaking and audacious range of cosmetics to the colour-starved post-war Dior customer.

In April 1953, the designer and his childhood friend from Granville, Serge Heftler-Louiche, launched *Rouge Dior* lipstick and added a daring collection of eight lipsticks ranging from a striking bright red to a pure orange.

And in 1968 Dior called upon its newly appointed Creative Director, Serge Lutens, to create a complete range of make-up colours for the house. Baptised *Explosion de Couleurs,* this new collection included a selection of 40 different shades of lipstick alone. The launch of the new line would be accompanied by a photographic campaign by Guy Bourdin in which Lutens painted audaciously vivid rainbows of prismatic colours – blue, green and yellow – across a model's face and immediately created another revolution at Dior, this time in cosmetics and in colour. As Peter Philips states emphatically, "Dior is all about colour!"

Dior's own childhood memories of the seaside town of Granville, where he grew up beneath the pebbledash façade of his parents' home "Les Rhumbs", are clouded with the colour of its low, leaden sky, the heavy swell of its sea, and the cliffs and rocky outcrops of its coastline. It created a "grisaille" of nuance and reminiscence, over which the designer would soon start to layer his cherished colours in a glaze of demonstrative reds and poetic florals to complement his monochromatic past. Grey was the base note, the undercoat over which he then added colour.

Although he claimed that his colour palette was not influenced by any particular philosophical or scientific theory, Dior had a wide knowledge of art history and was particularly aware of the contemporary art movements that had revolutionised the use of colour in the latter part of the 19th century and the beginning of the 20th: the Fauvist work of the Blaue Reiter group; the *plein-air* paintings of his beloved Impressionists who had flocked to Normandy, where the young Christian spent his idyllic childhood; the Post-Impressionism of Vincent van Gogh; the Expressionism of Otto Dix and Max Beckmann; and the Cubist abstraction of Georges Braque, Fernand Léger and Pablo Picasso.

The Dior family bought their property "Les Rhumbs" in 1905. The house had been built by a boat owner in the *style balnéaire* (seaside style) of the time, with its walls "roughcast in a very soft pink, mixed with grey gravelling", and the designer remained attached to these two colours all his life.

Christian Dior also admired the pink hues of the 18th-century Rococo painter Nicolas Lancret in works like *Mademoiselle de Camargo Dancing* (1730) and *Fête in a Wood* (1722), which perfectly evoke the tentative flowering of a woman's sensuality – a sentiment that intrigued not only 18th-century French Regency society, but also the designer, who took inspiration for the longer New Look silhouette from an ideal he fashioned of a woman as a flower: the "Corolle" silhouette opened up the skirt-like petals and revolutionised his new line. The fashion show notes of the time were full of references and information about "*rose*" in all its different variations, both as a colour and as a flower. Every description of shade and nuance bore witness to the designer's acute sense of colour and his understanding of composition, which had been influenced and informed by this childhood spent in Normandy. Dior explored all the possibilities of the floral landscape in his colours, prints and embroideries.

The colour pink continues to inspire designers at Dior and takes on new meaning particularly for Maria Grazia Chiuri who describes the colour as "the new bastion of modern feminism: it celebrates diversity, promotes its values and embraces being a stereotype that it can twist to its own advantage".

Dior's own life was indeed a "*vie en rose*", and when Edith Piaf, who wore both his clothes and his lipsticks, came to the designer's salon in 1955 the colour was already as identifiable with the house as the more sensual red of *Rouge Dior* or the modest tones of the salon interiors described by the designer as "my dear Louis XVI style with its *Trianon grey* and white shades". Dior's use of this palette immediately caught the attention of both clients and the press. Few designers had succeeded in identifying themselves with particular colours in such an effective and lasting way. They had become the triumphant colour code of his New Look and its unique revolution in luxury.

As he did all his signature shades, Dior the designer extolled the virtues of the colour grey in his 1954 *Little Dictionary of Fashion*, calling it "the most convenient, useful and elegant neutral colour". It was for him also the colour of Paris and he celebrated its shifting nuances on the catwalk with models wearing looks described variously in the show notes as "*gris – remparts*" (ramparts grey), "*gris phalène*" (moth grey), "*gris soupir*" (sigh grey), "*tons de ciel nacré*" (pearly sky tones), "*les gris petits jours*" (twilight greys) and "*ton du foyer, le gris hivernal*" (hearth colour, winter grey).

Coupled with grey, white was indeed the colour of the *grand salon blanc* at Dior's headquarters at 30 Avenue Montaigne, but also of the "French provincial" interiors of the Moulin du Coudret, Dior's country retreat in Milly-la-Forêt outside Paris. For Spring-Summer 1954, Dior dedicated a whole collection to lily-of-the-valley – considered the flower of purity and innocence and often associated with the Virgin Mary – as he created the *Muguet* line in their honour. The colour also evokes the slender curves of René Gruau's elegant swan fashioned for the first Dior fragrance *Miss Dior,* while in the Dior atelier the armchairs were upholstered in white and models drifted furtively through the salons swathed in spectral sheets to shield the secret of new designs from prying eyes. Today Maria Grazia Chiuri associates the colour "with the optimism generated by science and the positivist approach to reality".

If Dior often chose to wear white in his private life, black was the official colour of his public image. It is the colour of the elegant dandy and of the designer's favourite

1920s German Expressionist films including Robert Wiene's *The Cabinet of Dr Caligari* (1922) and Louise Brooks's raven bob in G.W. Pabst's *Pandora's Box* (1929). It is the singer Juliette Gréco, whom Dior described as "the perfect incarnation of a youth in tapered leg trousers and sweater black as cave rats", and it is the chic simplicity of Edith Piaf. It is Jean Cocteau's stark, sensual neoclassical line drawings, and Christian Dior's own "*petites gravures*" (little engravings) conceived at his country retreat in Provence, the Château de la Colle Noire, where the dark mass of the surrounding cypress-covered hills afforded him peace and sanctuary.

For Christian Dior, black could not live without white: they are the yin and the yang and exist in the contrasted jacket and skirt of a suit, or in the signature houndstooth motif that was both woven into cloth and printed on accessories and packaging.

In his *Little Dictionary of Fashion,* Dior calls the colour black "the most popular and the most convenient and the most elegant of all colours", and it was the source of constant inspiration for the designer. Like the painter Matisse, he used black as "a force, a ballast". It exemplified the simplicity, naturalness and distinction of silhouettes whose names echo their tone, like *Dame en Noir* (Autumn-Winter 1947), *Panthère Noire* (Autumn-Winter 1952) and *Collet Noir* (Autumn-Winter 1950). It is the colour of *Soirée de Paris* – the Autumn-Winter 1955 couture silhouette that the ebony-haired model Dovima incarnated in the celebrated 1955 portrait by Richard Avedon – and of the sombre eyes and delineated lips of Victoire Doutreleau as she glided like a dark angel through the grey and white showroom at 30 Avenue Montaigne.

For the designer, the only rival to black was navy blue, as the two colours shared similar enigmatic qualities. Dior's blue is the blue of elegance, of mystery. It is the blue of *Junon,* created in 1949, a dress whose voluminous grey tulle skirt and deep décolleté are entirely scattered with dark-blue embroidered sequins. It is "*la cliente bleue*" ("the blue customer"), a certain Mrs Firestone of Newport, Rhode Island, who only purchased gowns of royal blue from the designer. It is the court at Versailles and the delicate blue of Marie Antoinette's *toilette*, which the designer celebrated in his Spring-Summer 1955 collection. It is the blue of nostalgia for an idealised vision of an 18th-century society of balls and parties, and of the opera, where, before the French Revolution, all the chairs and tapestries were a heavenly blue.

Dior could have written not only a book on the colour black as he suggests in his celebrated *Little Dictionary of Fashion*, but on all the colours of the spectrum, such was his sensitivity for the subject. He had acquired this acute perception not only through a childhood appreciation of nature but also through an acquaintance with and knowledge of some of the leading artists of his generation.

In 1927, while still a young man and part of a social network of artists, writers and aesthetes in Paris – including his great friend and "brother" the painter and stage designer Christian "Bébé" Bérard, the poet and filmmaker Jean Cocteau, and the art patron and hostess Countess Marie-Laure de Noailles – Dior had opened a gallery in partnership with his friend Jacques Bonjean, on the Rue La Boétie. The gallery exhibited many of the most influential and celebrated artists of the first half of the 20th century, including Giorgio de Chirico, Maurice Utrillo, Salvador Dalí, Raoul Dufy, Marie Laurencin, Fernand Léger, Jean Lurçat, Pablo Picasso, Ossip Zadkine, Georges Braque and Aristide Maillol. Dior's relationship with these artists would be central to his own exploration of colour and his work as a designer.

Maria Grazia Chiuri continues to champion this spirit in her work today when for her Spring-Summer 2018 collection, she looked to the work of the French artist Niki de Saint Phalle whose vibrant colours she grafted onto a white base to create what she describes as "graphic patterns with stripes or polka dots, slogans stating the Dior aesthetic's name and new personality". In her Pre-Fall 2019 collection, the designer also used unexpected watercolour hues inspired by painter Sonia Delaunay's palette which in her words were "unconventional for me as well as for the House".

Colour and its visible spectrum of shifting shades continues to exercise a unique fascination over the minds and eyes of artists and scientists alike, from the earliest cave painters 75,000 years ago, who daubed primitive signs and symbols on stone walls with ancient minerals like charcoal and red ochre, to today's nanotechnologists, digital explorers and design visionaries like Maria Grazia Chiuri and Peter Philips.

Together their own innovative approach to the art and language of colour heralds another era for Dior and the Dior Woman, always evolving, always changing, always reinventing herself.

THE GLOBETROTTER WHO MADE THE WORLD LOOK NEW

QUEENNIE YANG

Christian Dior's family had once hoped he would become a diplomat. Due to his passion for art and fashion, he instead eventually became one of the greatest designers in history. But still, he travelled around the globe, leaving his footprints from Europe to South America, and conquered the world with his stunning creations. He had become the prime ambassador of French style.

His desire to travel was deep-rooted. England was the first country he visited in his youth. He spent his twenties travelling from Athens to Leningrad and Istanbul, as well as a year in the Balearic Islands. From his debut collection, he took inspiration from the cultures he had observed, paying tribute to their architecture, landscapes and art by naming his dresses after the places that inspired him.

As his fame grew, his fashion house expanded to a global realm, spanning from the Americas to Japan and from Mexico to Australia, by opening boutiques and licensing his designs. This gave him more styles to draw on, and increased his opportunities to travel. He was one of the first international designers to appreciate the importance of the global market and be truly loved by customers all over the world. His travels fed into his creativity as reimaginings of what he had seen, and these were later inherited by his successors. Through his interpretations, these different cultures made their entrance into the international pantheon of fashion.

The Far East was always a creative wonderland for him. Growing up in Granville, the interiors of his family home were decorated with japonaiserie. "Those interpretations of Utamaro and Hokusai were my Sistine Chapel. I used to spend hours just gazing at them," he recalls in his memoirs. The exquisite embroidered silks left an unforgettable impression on him and were evoked regularly throughout his collections. He often chose rich silks with traditional Japanese motifs. In 1953, he created an afternoon ensemble called *Jardin Japonais*, with a pattern of cherry blossoms, flowers and birds; in 1954, he designed a kimono-inspired costume for the ballerina Margot Fonteyn in the ballet *Entrée Japonaise*. He sent the House of Dior models to Tokyo to show off his haute-couture collection there, and designed three dresses for Empress Michiko's civil wedding.

In his autobiography, Dior also often describes the intricacy of Asian craft and design and his preference for fabrics and designs from India and Japan, which would be blended into his modern aesthetics. He was deeply fascinated by India's techniques of embellishment and incorporated traditional embroidery into his evening dresses that later made him one of the greatest ballgown makers of all time. The models Anne Gunning and Barbara Mullen dressed in Dior and wandering around Mysore City and New Delhi were captured in photographs by Norman Parkinson and featured in British *Vogue*, displaying an enchanting fusion of beauty.

China was another everlasting fascination for Dior, although he had never been to China in his life. In his first collection, alongside the famous New Look, he also created an overcoat baptised *Shanghai* and paired it with a Tonkinese hat. Then he paid tribute to this ancient and mystical country via creations such as *Pékin*, *Shanghai*, *Chinoiseries*, *Nuit de Chine*, *Hong Kong* and *Bleu de Chine*. The beauty of Chinese calligraphy and the traditional dress known as the qipao were also celebrated by him. He used ideograms for the print for a dress in the *Verticale* line in 1950 and reinterpreted qipao in vivid pink, for the ensemble *Surprise* from Autumn-Winter 1955. His creation of "Chinese" dresses was singled out by *Vogue*, which praised the designer for choosing themes that were Oriental in origin but contemporary Occidental in effect.

On the other side of the Pacific, the Americas, especially Central and South America, were another exciting challenge for Dior. He travelled to the USA in 1948, and in the same year the owners of the Neiman Marcus department store presented him with their Fashion Award in Dallas. The couturier then set off on a triumphant tour of the American continent. In 1953, he received a warm welcome in Caracas, Venezuela. The following year, he toured Bogota, Mexico City, Panama City, Puerto Rico, Kingston, Curaçao and Havana with a "travelling fashion show". In Havana he paid multiple visits to the luxury boutique El Encanto, and in Mexico City he visited the department store El Palacio de Hierro. Later, these two became the ideal places to showcase his creations. His signature A-line skirts are reminiscent of those worn by Mexico's Tehuana women, and he named several of his dresses after Mexico, with dot print, floral decorations and dazzling golden embroideries.

In 1951, for the Venetian Ball of the Century, Dior dressed Daisy Fellowes, the French socialite and Paris editor of American *Harper's Bazaar*, as a fierce Queen of Africa with bold leopard print and feather headdress. Apparently, the idea of Africa and Ancient Egypt was another fascination of his. He would bestow names such as *Cleopatra* and *Ramses* on his dresses.

Like Christian Dior himself, all of his successors have been enthusiasts of travel and other cultures.

Although Yves Saint Laurent, Dior's great protégé, did not run the House of Dior for very long, he later demonstrated his love for China, Peru, Morocco and wider exotic cultures at his eponymous brand.

Marc Bohan had the longest tenure as Creative Director. In the Autumn-Winter 1962 *Flèche* line, Bohan applied globe-trotting names from *Air France* to *Paris Tokyo* and from *Soirée à Bangkok* to *Saint-Sylvestre à Rio* to honour this great tradition. And later in the Spring-Summer 1965 collection, he reinterpreted Asian culture into colourful Hindustan prints and kaftans. Africa inspired him to create the Autumn-Winter 1964 collection with tube dresses and the Spring-Summer 1967 collection with "African style" dresses and "Safari" suits. Bohan did not take a clichéd approach but abstracted the palette and decorative motifs of Africa into very simple and modern silhouettes. His Spring-Summer 1966 collection did likewise with Mexican sources. He poured his passion for extravaganza into the two collections of 1968, which featured a profusion of luxurious embroidery, gold accessories and gowns to represent the glory of the Byzantine era. In 1962, he staged the brand's first fashion show in Mumbai and designed its first saris. Twenty-seven years later, his Spring-Summer 1989 *The Indian Year* collection (*Pyramid* and *Sari* lines) made for a perfect ending to his tenure at Dior with its bright pinks, yellows and oranges.

The Italian designer Gianfranco Ferré made many extended trips to India and in his final year at Christian Dior also paid homage to that country: his Autumn-Winter 1996 collection, titled *Indian Passion*, was inspired by its colours and treasures and recreated its embroideries, saris and turbans. Ferré occasionally put a British accent into his designs as well, which likewise shows a deep connection with the firm's founder. Edwardian high society and Cecil Beaton's costumes inspired his Autumn-Winter 1989 *Ascot – Cecil Beaton* collection; and Shakespeare's *A Midsummer Night's Dream* found its interpretation in the feather-light Spring-Summer 1990 collection. And of course,

Ferré revisited his mother country for Autumn-Winter 1992, with the *In the Secret of a Venetian Winter* collection, to remind people of this city's splendour in the 18th century.

The rebellious Englishman John Galliano is probably the wildest character to have been in charge of the House of Dior. His unlimited imagination melded seamlessly with his travels, and he went on to create countless surreal journeys and iconic looks for the catwalk. In his world, travel was not limited by time or space. The Maasai tribe encountered chinoiserie, while Shanghai pin-up girls' qipao showcased hyper-femininity with a twist of Hollywood stardust. The Marchesa Casati, Mata Hari and Madame Butterfly were his imaginary travelling companions. He took the "Diorient Express" with both Princess Pocahontas and Wallis Simpson, and invited Empress Sissi of Austria on a trip across central Europe to Istanbul. From Russia to Mongolia, he not only spotted the tribal motifs but also Russian avant-garde art and the Chinese Red Army as inspirations. Hokusai's famous wave print that once decorated Christian Dior's Granville villa was recreated by Galliano in his Spring-Summer 2007 Haute Couture collection, with fabric origami and a stunning coat. And for the marvellous Spring-Summer 2004 collection, inspired by his tour of Egypt, the signature *H* line was decorated by flamboyant and gilded shades, from Nefertiti to King Tut, literally coming alive from the Valley of the Kings.

Compared to Galliano, the Belgian Raf Simons seems more rational and mellow, but no doubt he is also an internationalist just like the firm's founder and his predecessors. A modernist, he prefers to plug into a wider approach of world culture instead of using direct references. By reimagining Andy Warhol and inviting Sterling Ruby to collaborate, he injected the most exciting parts of American arts into Dior. In his Autumn-Winter 2013 *World Couture* collection, he represented four distinctive realms: "Europe", focused on the firm's French heritage; "The Americas", which was all about dynamic bold spirit; "Asia", which featured "clothing full of balance, tradition and purity"; and "Africa", showing "freedom, playfulness and effortless creativity", along with a tribute to Maasai style. These codes were subtle yet strong. But only when Simons transported us to the "megalopolis" in Tokyo for his Pre-Fall 2015 show did we finally realise how deeply he was in love with Japanese culture. The kawaii style and avant-garde architecture inspirations both shone out at the show. His vision had a futuristic gloss, where fantasy and reality co-existed. His destination was the future.

When Maria Grazia Chiuri came to Dior, she brought her Italian romanticism with her, but more importantly, femininity and feminism. The first woman in charge of the firm also faces the challenge of defining the line between cultural appropriation and appreciation. These days, the Parisian House of Dior literally travels around the world through its Resort, Pre-Fall and even Haute Couture shows, and Chiuri uses these opportunities to honour different cultures and inspire women globally. For the 2018 Resort collection, which was shown in an open-space reserve in Calabasas, USA, she connected the House founder's "Lascaux" collection inspired by the ancient cave paintings discovered in south-western France to California's wild side. From new cowboy hats to plaid and tassels, Western elements were everywhere. For the Resort collection the following year, Chiuri was touched and inspired by Mexican female riders of the *escaramuzas* who were forbidden from participating in all-male *charreada* equestrian events, and so invited an eight-woman band of female rodeo stars to the catwalk and translated this spirit into glamorous Mexican patterns, Chantilly lace and other inspirations from the Magic Realism of Latin American literature. In many other collections, Chiuri pays tribute to Christian Dior by injecting his iconic designs into modernised forms. For instance, the 1953 *Jardin Japonais* collection was recreated with diving birds and trailing cherry blossoms in the Spring-Summer 2017 Haute Couture collection shown in Tokyo as dedicated new designs. She also paid homage to the House founder's lucky colour red with her Shanghai Couture show. And a 1953 illustration by Albert Decaris depicting Christian Dior's travels guided her to develop the Autumn-Winter 2018 Haute Couture collection. The forward-thinking couturier once said in his autobiography that "a complete collection should address all types of women in all countries"; Chiuri is making it happen.

Although Christian Dior himself is no longer with us, the French fashion house and his successors are far from reaching an end of their voyage of discovery. They will keep changing the European fashion landscape to become more inclusive and modern, and celebrating people and cultures from all around the world.

DIOR AFTER DIOR

HUANG HUNG

When Christian Dior suddenly passed away in October 1957, both the man and the fashion house he had created were at their zenith. In 10 years, Christian Dior had defined aesthetic standards for the coming century: ever since his New Look became an instant sensation in 1947, the name Dior has been synonymous with feminine beauty.

Over 60 years after his death, the legend of Dior lives on. This is a testament not only to the greatness of Christian Dior as a couturier, but also to all the creative directors who followed him and enabled the House of Dior to remain the leading couture house throughout the decades. Fittingly, then, the exhibition *Christian Dior Designer of Dreams* is not only about Dior the man, and the fashion house he founded, but also about its six outstanding creative directors – Yves Saint Laurent (1958–60), Marc Bohan (1960–89), Gianfranco Ferré (1989–96), John Galliano (1996–2011), Raf Simons (2012–15) and Maria Grazia Chiuri (2016–present) – each of whom continued his legacy while interpreting it for their own times.

The word “heritage” is often connected with the verb “to protect”. In Chinese culture, heritage is protected through copying: the greater the likeness to the master’s work, the better. To deviate from the master’s work is sacrilege. For instance, students of classical Chinese painting have been copying the same paint strokes from the book *The Garden of Jianzi* since the Qing Dynasty (1636–1912).

Christian Dior Designer of Dreams will be enlightening to the Chinese audience: the exhibition clearly demonstrates that creativity is as important as protection in safeguarding a heritage. These six great designers have not only protected the aesthetics and DNA of the Dior brand, they have also used their own creativity to infuse new life into it, ensuring its continuing relevance. What this beautiful volume therefore presents is an account of the building of a brand, and an unforgettable journey of creation.

The Builders

Marc Bohan, probably the least known in China of the six designers, deserves credit for consolidating the reputation of the House of Dior as well as making its signature looks a standard for beauty. When he became Creative Director, Bohan promised: "Everything I design will be in the true Dior tradition." He later went on to say: "As for me, I wouldn't know how to 'do' Dior. I admire him hugely, but I never worked with him, so I was never tempted to copy him. I endeavoured to retain all that forged his reputation, which is to say the elegance and distinction of his work and its special charm, but to interpret it through my own style." These two seemingly contradictory statements define the essence of Bohan's success at Dior: he interpreted all the signature Dior looks, but added a twist of 1960s and 1970s modernity to them. Take for example the *Miss Dior* dress – a name Christian Dior had often repeated in different collections. Bohan relaxed the strictness of the dress and made it more comfortable. His interpretation created probably the biggest celebrity clothing clash of the 20th century when two movie megastars both wore *Miss Dior* on the same day at the Moscow Film Festival in 1961.

Bohan's very flattering *Gamin* suit and *Slim* series paved the way for his launch of Dior Ready-to-Wear in 1967. The fact that today we can walk into a Dior store and pick our favourite dress is largely thanks to him.

Bohan's successor Gianfranco Ferré was not given the benefit of the doubt by the fashion press when he took up the position of Creative Director at Dior: its writers were largely sceptical of an Italian as the creative head of the most venerable French fashion house. Ferré came from ready-to-wear but was always fascinated by haute couture. In fact, before arriving at Dior, he had had his own couture line in Rome, which he had closed down to focus fully on his job at Dior. Ferré and Christian Dior shared two similar passions. One was opulence, which was particularly "in" in the 1980s. The second was architecture. Ferré was able to deconstruct architecture and reconstruct details into couture, something Christian Dior had done in his time as well. The *Roman Column* evening gown by Ferré, which is featured in the exhibition, is an exquisite example of the fusion between architecture and fashion.

In terms of working style, however, Ferré could not have been more different from Christian Dior. He brought more openness and made the atelier more democratic. He was a chatty man, so he rearranged the workspace to improve communication between different workstations. In a word, he modernised the way couture was made.

The Rebels

Yves Saint Laurent and John Galliano were both rebels during their time at Dior. The two have completely different personalities: Saint Laurent was an introvert, like Dior himself, while Galliano is an extrovert. Both understood shock value and used it cleverly to bring Dior to the cutting edge of fashion.

In some ways, Christian Dior was a rebel himself as well. During the frugal post-Second World War years, it was considered a sin to waste so much fabric on a single dress.

In every rebel, there is a heart that wants to please. Both Saint Laurent and Galliano worshipped Christian Dior. At the press announcement of his appointment as creative director, Saint Laurent said: "I cannot describe to you everything I feel, it would take too long: sadness, anguish, at the same time joy, pride, the fear of not succeeding." For his part, Galliano, in an interview in his early days at Dior, said with almost childlike simplicity: "I just want to make things that Christian Dior would like."

Both designers created the most controversial collections in the history of the House of Dior. Saint Laurent was very much into the Beat Generation, and was the first designer to show a leather jacket inspired by motorcycle attire in a couture collection. This garment is in the exhibition, where you can see that despite its inspiration and the controversy that it created at the time, its elegance is very much in the Dior tradition.

Saint Laurent's tenure at Dior was cut short in 1960 when he was conscripted to serve in the French army during the Algerian War of Independence.

Galliano had a longer tenure at Dior, and created sensation after sensation. Every season, Dior under Galliano was the talk of the town. His year 2000 collection inspired by hobos, for instance, boasts an opulence and elegance that was very much in the Dior tradition and fitted the aesthetics of the bourgeoisie; yet irony and sarcasm are also readily apparent in it. The "hobo" pieces from this collection sparked some of the greatest controversy in fashion history. Galliano's ability to create drama with fashion can be attributed to the fact that he had worked in theatre. For him, fashion is much more than clothes: it is a language which can be used to express yourself, create emotion and even write social criticism.

Galliano was also very skilled at his craft. He was known for his bias cuts – a cut which allows the fabric to be more elastic and hence drape over the body in a very shapely manner. In the exhibition, there is a navy-blue satin evening gown that Galliano made for Princess Diana in 1996. She wore it at that year's Met Gala and was immediately coined "Sexy Di" by the press. To be so skilled at playing with the language of fashion is truly the work of a genius.

The Philosophers

When Galliano left Dior, it was the end of the era of fashion designers as superstars. Both Raf Simons and Maria Grazia Chiuri are intellectual designers. Both spent huge amounts of time studying the Dior archive and drew endless inspiration there. Their first collections at Dior were tributes to Christian Dior's own first collection.

Raf Simons shared Christian Dior's passion for art, and was able to bring contemporary art into the language of Dior. His debut collection was a stunning tour de force of all the signature Dior elements but with a surprise. To bring Dior's love of gardens and the *femme-fleur* into the mix, the walls and ceilings of the show space were covered with fresh flowers. His first look, a black trouser suit with the classic *Bar* jacket, deftly created a permanent image of the Dior woman in the 21st century.

Maria Grazia Chiuri is the first woman to be named as the Creative Director of the House of Dior. She has not only proved that couture is about more than just evening gowns, but also infused her own philosophical thoughts on feminism into each collection, bringing to the forefront of fashion ideas such as "we all should be feminists" and "the power of the sisterhood". Trained as a fashion designer and with extensive prior experience at both Fendi and Valentino, Chiuri is extremely skilled. Her ability to create weightless opulence is proof of her mastery and creativity in working with fabric. While true to Dior's DNA, Chiuri firmly points fashion to the future.

Christian Dior Designer of Dreams is an exhibition of many stories. Writing as a journalist who is longing to see the emergence of a global Chinese brand, this exhibition is truly special. For it tells the story of not only this iconic brand, but also the endless creativity that has preserved it and made it relevant throughout the years. For me, this exhibition is a glamorous classroom that offers lessons in brand building through telling the story of a legend that is ongoing.

FAIRY NIGHTS

JUSTINE PICARDIE

From the moment that Christian Dior launched his first collection in February 1947, it became clear that romanticism was at the heart of his creative aesthetic, epitomised by exquisitely beautiful ballgowns and eveningwear. The femininity of these flowing dresses reflected those worn by his mother Madeleine, during his childhood in the Belle Époque, a golden age of peaceful prosperity before the onslaught of the First World War. By the time he had started designing clothes in the late 1930s, another catastrophic war was already looming; and having been employed at Lucien Lelong's couture house in Paris during the early 1940s, Dior was well aware of how the German Occupation had reduced French fashion to a shadow of its former self.

In his memoir *Dior by Dior* (first published in 1956, in French), he wrote: "As a result of the war and uniforms, women still looked and dressed like Amazons. But I designed clothes for flower-like women, with rounded shoulders, full feminine busts, and hand-span waists above enormous spreading skirts." This graceful silhouette, using extravagant quantities of fabric, was described by Carmel Snow, the influential editor of *Harper's Bazaar*, as "such a new look". In truth, it only appeared revolutionary in the context of wartime textile rationing and post-war austerity; and one of the most popular evening dresses in Dior's first collection – *Soirée*, a tiered navy-blue silk taffeta gown with a low-cut bodice and ribbon shoulder straps – marked a return to Hollywood glamour. It therefore seemed a natural choice for the film star Rita Hayworth to wear to the Paris premiere of *Gilda* (in which she starred in her signature role as the ultimate femme fatale), thereby ensuring that legions of her American fans were dazzled by Dior's vision of elegance.

Each of his subsequent shows presented equally alluring designs: for example, *Fête*, a delicate pink floor-length dress adorned with a pleated bustle in the Spring-Summer 1948 collection; and the following season, *Virginie*, a white satin ballgown with a sky-blue faille sash. The joyous celebration of beauty continued to flourish: the *Miss Dior* evening dress in the Spring-Summer 1949 collection was an enchanting sartorial bouquet, blossoming with a thousand embroidered flowers.

Yet Christian Dior's concept of style was not confined to a youthful prettiness. If *Miss Dior* was inspired by his beloved younger sister Catherine, then the dramatic black velvet evening gown *Cygne Noir* (presented in the Autumn-Winter 1949 show) was redolent of the grown-up sophistication embodied by his muse, Mizza Bricard. Indeed, this was a virtuoso collection notable for its graceful panache, with the subtle pale-grey beaded tulle of the *Venus* gown appearing alongside the cascading midnight-blue sequins of Dior's masterpiece, the magnificent *Junon*. These were not dresses for ingénues, but for goddesses ...

No wonder, then, that Marlene Dietrich would only be seen in Dior – as she made clear to Alfred Hitchcock before agreeing to accept the lead role in his film *Stage Fright* in 1950: "No Dior, no Dietrich!" A regular guest at Christian Dior's shows, and at his home in the countryside, she wore his dramatic designs (including *Cygne Noir*) with confident aplomb.

Dior's association with royalty was just as powerful. When the young Princess Margaret travelled to Paris in the spring of 1949, on her first European tour, she visited the couturier at his salon on Avenue Montaigne, to order several dresses; and the following year, she was an appreciative guest at a private Dior show at the French embassy in London, staged for a small audience including her mother and several other members of the Royal Family. In 1951, Princess Margaret chose a fairytale white Dior ballgown to wear for her 21st birthday, its organza skirt and bodice embroidered with gold and mother-of-pearl. This was, she declared, "my favourite dress of all", and she was immortalised wearing it in a famous official portrait by Cecil Beaton.

Christian Dior's status as the preeminent couturier of choice for royalty and Hollywood was endorsed by Grace Kelly; for having embodied the ideal of the American heroine in her film roles, and on the red carpet (where she looked poised to perfection in Dior couture), she was then transformed into the Princess of Monaco. The supremely elegant white satin Dior gown that she wore for the ball celebrating her engagement to

Prince Rainier in January 1956 became one of the most iconic looks of the era, and she remained loyal to the house of Dior for the rest of her life.

"A ballgown is a dream, and it must make you a dream," declared Christian Dior, and his genius for creating fantasy and delight reached new heights at what has been dubbed "the party of the century" – held at the Palazzo Labia in Venice in September 1951. The host was Don Carlos de Beistegui, a fabulously rich playboy who had lavished close to a million dollars on restoring the palazzo after he bought it in 1948; and it was to become the stage for 1,500 of his guests, described by *Life* magazine as "the world's most blue-blooded and/or richest inhabitants". Christian Dior was amongst those invited, and he also designed ballgowns for many of the guests, including Daisy Fellowes (formerly the Paris correspondent for *Harper's Bazaar*, and a socialite famed for her cruel wit, devastating chic and immense wealth). According to another of the guests, David Herbert, "Daisy's dress was made of yards of champagne-coloured chiffon with a subtle print of leopard skin". As for Dior himself, he recalled the Beistegui ball in his memoir: "This was the most marvellous spectacle which I have ever seen, or ever shall see. The splendour of the costumes rivalled the splendid attire of the figures in the Tiepolo frescoes on the walls. (...) The magic of a summer's night in Italy held us in its eternal spell and put us outside time."

It therefore seems fitting that Dior's current Creative Director, Maria Grazia Chiuri, should have made such equally striking costumes for the memorable Venetian ball held at the Palazzo Labia in May 2019, creating ethereal gowns worn by guests including Karlie Kloss and Sienna Miller. On that night – as always – Chiuri celebrated the legacy of Dior, with her intuitive understanding that happiness and enchantment can be threaded through an evening dress. "In the world today *haute couture* is one of the last repositories of the marvellous," wrote Christian Dior in the final page of his memoir, "and the *couturiers* the last possessors of the wand of Cinderella's Fairy Godmother." Fortunately, Dior's wand has been passed down through the years, and remains as magical today as it was when he first wielded it, more than seven decades ago ...

本书为 2020 年 7 月 28 于上海龙美术馆举办的“克里斯汀·迪奥，梦之设计师”展览目录。

首先感谢酩悦轩尼诗路易威登集团的董事长兼首席执行官贝尔纳·阿尔诺，迪奥时装公司董事长兼首席执行官皮特罗·贝卡里，迪奥香氛公司董事长兼首席执行官洛朗·克莱特曼，迪奥时装及香氛公司中国区总裁埃尔维·佩罗，迪奥时装公司负责国际沟通宣传的副总裁奥利维耶·比亚罗伯，迪奥香氛公司负责国际沟通宣传的副总裁杰罗姆·皮里对本次展览的大力支持。

衷心感谢龙美术馆馆长王薇，维多利亚和阿尔伯特博物馆现代时装及纺织品部及本次龙美术馆迪奥展览策展人奥丽悦·库仑，迪奥时装公司文化项目主任艾莲娜·斯塔克曼，迪奥时装及香氛公司中国区资深副总裁徐茜。

同样感谢所有为本书出版做出贡献的人，尤其是摄影师拉齐兹·哈曼尼、郎港澳；撰稿人娜迪亚·阿尔贝蒂尼、洪晃、克蕾莉·艾泽、菊丝婷·皮卡迪、杰瑞·斯塔福德、杨忆非、卡特琳·泽彤。

本书装帧设计为 Adulte Adulte 工作室的弗洛朗·佛里、达尼埃尔·里贝罗。

迪奥公司的杰罗姆·郭蒂埃、董晴、杰拉·舍瓦利耶、塞巴斯蒂安·克里瓦、杰罗姆·威尔布拉克、索契·法夫、索兰·奥蕾雅拉米、卡米尔·比度兹、维多利亚·布拉其、阿梅丽·伯撒、塞西尔·莎木爱卡那、奥利维耶·科龙巴、大卫·达席尔瓦、芭芭拉·若夫麦蕾、菊丝婷·拉吉、尼古拉·罗、宝霖·帕亚、索伦·罗杰曼、佩玲·谢莱、乔安娜·托斯塔、佩玲·谢莱、乔安娜·托斯塔、杰尼芙·瓦蓝、菲利普·勒姆、玛丽·奥德朗、安夏洛特·梅西耶、斯特凡妮·培良；

迪奥香氛公司的弗雷德里克·布德里耶、文森·乐海、桑德琳·达美 - 布鲁；

卡西出版社的卡特琳娜·伯尼法希、瓦内莎·布隆戴尔，以及 mot.tiff inside 工作室的亚历山大·维希内夫斯基、卡罗琳娜·加吕伏、克莱尔·安托万、胡小力等都为本书出版做出了宝贵贡献。

This catalogue has been published to coincide with the *Christian Dior Designer of Dreams* exhibition at the Long Museum, 28 July 2020.

The project was made possible thanks to support from Bernard Arnault, LVMH Chairman and Chief Executive Officer, Pietro Beccari, Chairman and CEO of Christian Dior Couture, Laurent Kleitman, President and Chief Executive Officer at Christian Dior Parfums, Hervé Perrot, President of Christian Dior Couture and Christian Dior Parfums in China, Olivier Bialobos, Chief Communication Officer at Christian Dior Couture, and Jérôme Pulis, International Communication Director at Christian Dior Parfums.

We would like to express our sincere thanks and gratitude to Wang Wei, Director of the Long Museum, Oriole Cullen, curator of Modern Textiles and Fashion at the Victoria and Albert Museum and curator of the Dior exhibition at the Long Museum, Hélène Starkman, Cultural Projects Manager at Christian Dior Couture, and Lucy Xu, Senior Vice President Global Communication at Christian Dior Couture and Christian Dior Parfums in China.

We also would like to thank all those who have made the publication possible, and particularly:

Photographers Laziz Hamani and Gangao Lang, authors Nadia Albertini, Huang Hung, Clélie Haize, Justine Picardie, Jerry Stafford, Queennie Yang and Catherine Zeitoun.

Artistic directors of the catalogue Florent Faurie and Daniel Ribeiro at Adulte Adulte.

Exhibition designer Bertrand Houdin at Anamorphée.

Jérôme Gautier, Noemie Dong, Gérald Chevalier, Sébastien Clivaz, Jérôme Verbrackel, Soizic Pfaff, and also Marie Audran, Solène Auréal-Lamy, Camille Bidouze, Victoria Blazy, Amélie Bossard, Cécile Chamouard-Aykanat, Olivier Colombard, David Da Silva, Barbara Jeauffroy-Mairet, Justine Lasgi, Philippe Le Moult, Nicolas Lor, Anne-Charlotte Mercier, Pauline Paillard, Solenn Roggeman, Perrine Scherrer, Joana Tosta, Jennifer Walheim, and Stéphanie Pélian at Christian Dior Couture.

Frédéric Bourdelier, Vincent Leret and Sandrine Damay-Bleu at Christian Dior Parfums.

Catherine Bonifassi and Vanessa Blondel at Cassi Edition. Sacha Vichnevski, Karolina Galluffo, Claire Antoine and Hu Xiaoli Violette at mot.tiff inside.

克里斯汀・迪奥，梦之设计师
Christian Dior Designer of Dreams

导言：
奥丽悦・库仑
Introduction:
Oriole Cullen

撰稿：
娜迪亚・阿尔贝蒂尼、洪晃、
克蕾莉・艾泽、菊丝婷・皮卡迪、
杰瑞・斯塔福德、杨忆非、卡特琳・泽彤

Texts:
Nadia Albertini, Huang Hung, Clélie Haize, Justine Picardie, Jerry Stafford, Queennie Yang, Catherine Zeitoun

摄影：
拉齐兹・哈曼尼、郎港澳

Photography:
Laziz Hamani, Gangao Lang

策划、装帧设计、执行编辑：
卡西出版社
卡特琳娜・伯尼法希、瓦内莎・布隆戴尔、阿比盖尔・格拉特、
坎迪斯・纪尧姆

弗洛朗・佛里、达尼埃尔・里贝罗、
罗恩・波特（Adulte Adulte 工作室）

Editorial Direction, Design and Production:
CASSI EDITION
Catherine Bonifassi, Vanessa Blondel, Abigail Grater, Candice Guillaume

Florent Faurie, Daniel Ribeiro and Loan Bottex for Adulte Adulte

中文翻译：
mot.tiff inside 工作室：
亚历山大・维希内夫斯基、
克莱尔・安托万、卡罗琳娜・加吕伏、
胡小力

Chinese Translation:
mot.tiff inside
Sacha Vichnevski, Claire Antoine, Karolina Galluffo, Xiaoli Hu